Applied Sport
Psychology

Applied Sport Psychology

A Case-based Approach

Editors

Brian Hemmings

and

Tim Holder

St Mary's University College, Twickenham, UK

WILEY-BLACKWELL

A John Wiley & Sons, Ltd., Publication

Library of Congress Cataloging-in-Publication Data

Applied sport psychology : a case-based approach / edited by Brian Hemmings and Tim Holder.
  p. cm.
Includes bibliographical references and index.
ISBN 978-0-470-72573-3 (cloth)
  1. Sports–Psychological aspects. 2. Psychology, Applied. I. Hemmings, Brian. II. Holder, Tim, 1939-
  GV706.4.A655 2009
  796′.01–dc22

                                                                                    2009005624

ISBN 978-0-470-72573-3 (HB)   ISBN 978-0-470-72574-0 (PB)

A catalogue record for this book is available from the British Library.

Set in 10.5/13.75 Minion by Laserwords Private Ltd, Chennai, India.

First Impression   2009

Brian

To my wife Kim and our three wonderful girls Harriet, Katie and Natalie – remember work hard, be busy and have fun!

To the rest of the Hemmings family, you are a great bunch and I am proud of you all.

Tim

To Ann – for always being patient, understanding and providing unreserved encouragement and support that enables me to function – without you none of it would be worth doing.

To Mum and Dad – for your unconditional love and support – I hope you realize how important you have always been.

Contents

Series Preface

One of the most astonishing cultural phenomena of the twentieth century has been the exponential growth in our knowledge and understanding of the importance of sport and exercise to human kind. At the beginning of that century, sport was principally a force for moral development, whilst strenuous exercise, though necessary to ensure military personnel were fit to engage in combat, was medically proscribed. The academic study of sport – what there was of it – was restricted largely to the history of the Olympic Games and philosophical arguments for the moral case for team games. A hundred years later, the picture is very different. 400 million people turn on their television sets to watch the Opening Ceremony of the Olympic Games and soccer's World Cup Final; millions of people jog, go to the gym, or work out in front of the television; and the academic study of sport embraces physics, chemistry, biology, biomechanics, physiology, psychology, politics, sociology, social anthropology, and business studies as well as history and philosophy. Over the last twenty years the number of degree courses in the academic study of sport and exercise has grown phenomenally, attracting students from a wide range of backgrounds. It is against this background that the new series *Wiley SportTexts* was conceived.

This new series provides a collection of textbooks in Sport and Exercise Science that is rooted in the student's practical experience of sport. Each book covers the theoretical foundations of the contributing disciplines from the natural, human, behavioural, and social sciences, and provides the theoretical, practical and conceptual tools needed for the rigorous academic study of sport. Individual texts focus on a specific learning stage from the various levels of undergraduate to postgraduate study.

The series adopts a student-centred, interactive, problem-solving approach to key issues, and encourages the student to develop autonomous learning strategies through self-assessment exercises. Each chapter begins with clear learning objectives and a concise summary of the key concepts covered. A glossary of important terms and symbols familiarises students with the language and conventions of the various academic communities studying sport. Worked examples and solutions to exercises together with a variety of formative and summative self-assessment tasks are also included, supported by key references in book, journal and electronic forms. The series will also have a dedicated website with specific information on individual titles, supplementary information for lecturers, important developments in the academic study of sport, and links to other sites of interest.

Eventually, it is intended that the series provides a complete coverage of the mainstream elements of taught under- and post-graduate degrees in the study of sport.

Tudor Hale, Jim Parry and **Roger Bartlett**

Preface

Over the past two decades, the practice of sport psychology has received far greater attention and exposure in the media due to its perceived importance in high-performance sport. The field of sport psychology has also gained increased professional recognition, whereby more career opportunities are now emerging. In turn, there is an increasing number of specialist sport and exercise psychology courses available at undergraduate and postgraduate level in the UK.

Furthermore, in 2004 the British Psychological Society (BPS) created a specialist division for sport and exercise psychology and chartered status is now available to suitably educated and trained individuals. At the time of writing there are currently 153 Chartered Sport (and Exercise) Psychologists registered with the BPS. Additionally, the British Association of Sport and Exercise Sciences (BASES) has an accreditation programme for individuals who wish to train as sport (and exercise) scientists and offer psychology support to athletes. Currently, there are 183 individuals who have accredited status. These numbers point to a discipline still in its infancy in the UK, with the potential for a huge growth in the knowledge and application of psychology to performance enhancement in sport.

This is not to suggest that sport psychologists solely work on improving performance, as many interventions address, in addition, the well-being and personal growth of the athlete. However, Anderson *et al.* (2002; *The Sport Psychologist*, *16*, 432–453) report that in a survey of BASES-accredited UK practitioners, the most frequently stated aim when working with athletes was to improve performance. Additionally, a review of the existing literature reveals a multitude of theory- and research-focused texts, with much less written on the practice elements of sport psychology in performance. This is surprising given that the very lure of sport psychology for many is their fascination with its 'real world' application. Furthermore, the experiences of practitioners are highly valuable, as applied knowledge is important in the shaping of future theory, research and practice.

The literature available to support the work of applied sport psychologists has been largely driven by academic rigour and scientific credibility. This leads from a logical identification that the practice of applied sport psychology should be derived from theory and the research evidence emerging from theoretical propositions. However, this book aims to show the broader reality of what applied practitioners do in real-world situations with athletes, focusing on *how* they work. This entails not only an acknowledgement of the knowledge base of theory and research informing the intervention with athletes, but also *practice-based* knowledge. This emphasizes the importance of experience and understanding of the consultancy situation (highly variable from sport to sport) and an understanding of how such experiential knowledge can be incorporated with the theory and research to attain successful working practices. In addition, the case studies within the book identify numerous examples where the theory and research evidence is being applied with athletes in a creative manner using flexible and adaptable approaches that work at the individual, team or organization level.

The case study approach has led to many discoveries in applied psychology, although its limitations for determining causal inferences are widely acknowledged by researchers. However, in addition to experimental research, further advances in applied sport psychology knowledge will no doubt require the use of non-experimental research methodologies, of which the case study is one such approach. Additionally, professional standards of practising sport psychologists in the UK are monitored and assessed by the above professional bodies, often through case study submissions on professional work undertaken with athletes. These case studies need to demonstrate high levels of intellectual and scientific rigour, and be evidence-based and contain reflective practice. Therefore, future practitioners and current students need to develop case study writing skills to understand the usefulness of case studies in developing applied work with athletes, teams and support staff in sport. This book aims to educate current and prospective sport psychologists about case study reporting, and simultaneously, to develop effective consulting skills for sport psychology practice.

The chapters contain unique case studies written by experienced and qualified practitioners to open the eyes of the reader to 'real world' issues and demonstrate the types of problems and challenges that a sport psychologist faces and is employed to resolve. It is worth noting that the book does not aim to offer exhaustive coverage of every issue that can be encountered in applied practice. There could be many more chapters on interventions with different groups, ages and populations of athletes, and with those in supportive roles. However, the book does cover a breadth of issues in an in-depth fashion to give the reader an insight into the *process* (e.g. building relationships) of support work, as well as the *content* of interventions. In doing so, the chapters involve a range of sports and athletes at different competitive levels of performance.

Brian Hemmings and **Tim Holder**

Authors' Biographies

Raphael Brandon
English Institute of Sport

Raphael Brandon is a UK Strength and Conditioning Association accredited coach. He has an undergraduate degree in psychology and an MSc in Sport Science. He joined the English Institute of Sport in 2003 and was the Lead Strength and Conditioning Coach for the Rugby Football Union Women (RFUW) from 2003 to 2006.

Sarah Cecil
English Institute of Sport

Sarah Cecil is the lead EIS psychologist for the London region. She has a BSc from St Andrew's University, an MSc from Exeter University and is a BPS chartered and BASES accredited sport psychologist. Sarah has worked extensively with a range of sports for over eight years and works currently with a number of Olympic sports, including athletics, fencing and shooting.

Iain Greenlees
School of Sport, Exercise and Health Sciences, University of Chichester

Iain Greenlees is a Reader in Sport Psychology at the University of Chichester. He gained his PhD on team confidence from the University of Southampton and is a BPS chartered and BASES accredited sport psychologist. Iain has worked with youth performers from a range of sports for over a decade.

Chris Harwood
School of Sport and Exercise Sciences, Loughborough University

Chris Harwood is Senior Lecturer in Sport Psychology at Loughborough University, where he completed his PhD in 1997. His research, teaching and training interests lie in the areas of achievement motivation, performance enhancement and psychosocial issues within youth sport environments. He is a BASES high-performance sport accredited psychologist and a BPS chartered psychologist. He has consulted widely with national governing bodies over the past 15 years, working within elite youth and professional tennis, football and cricket.

Brian Hemmings
Consultant Sport Psychologist and School of Human Sciences, St Mary's University College, Twickenham

Brian Hemmings is a Visiting Professor of Sport Psychology at St Mary's University College. He gained his PhD from the University of Southampton in 1998, and is a BPS chartered and BASES accredited consultant sport psychologist working full-time in private practice. Brian has worked extensively with a range of Olympic, professional and amateur sports for over 15 years and is currently actively involved with elite performers in international golf, cricket and motor racing.

Tim Holder
School of Human Sciences, St Mary's University College, Twickenham

Tim Holder is a Reader in Applied Sport Psychology at St Mary's University College. He is a BASES accredited sport and exercise psychologist and has a PhD from the University of Southampton. He has worked with a number of governing bodies since he started practising as a sport psychologist in 1989, including table tennis, swimming, squash, amateur boxing and sailing as well as consulting with individual performers from a range of sporting activities.

Jonathan Katz
Consultant Psychologist

Jonathan Katz has a PhD from Cranfield University and is a chartered (counselling and sport and exercise) psychologist and is BASES high-performance sport accredited. His work focuses on stress and performance: delivering performance under pressure with individuals, teams and organizations in sport and business, and he is an authority on counselling and effective communication systems. Jonathan has been the ParalympicsGB lead psychologist for the Athens, Turin and Beijing Paralympic Games.

Andrew Lane
School of Sport, Performing Arts and Leisure, University of Wolverhampton

Andrew Lane is a Professor of Sport Psychology at the University of Wolverhampton. He is a BPS chartered and BASES accredited sport and exercise psychologist and has authored more than 100 journal articles and has edited two books. His applied work has included athletes at the English Institute of Sport and the London Boxing Association, where he supported preparations for World Championship contests. A former amateur boxer, he remains active in sport as a runner and duathlete.

Caroline Marlow
Sport Performance, Assessment and Rehabilitation Centre, Roehampton University

Caroline Marlow is a Senior Lecturer in Sport Psychology at Roehampton University. She has a PhD from Brighton University and has been a BASES accredited sport and exercise psychologist for over 10 years. Caroline has worked with athletes from a range of sports and levels, and particularly specializes in providing psychological support to young athletes. She currently works with performers from cricket, badminton, gymnastics and bowling.

James Moore
English Institute of Sport Sheffield

James Moore is a chartered physiotherapist with a BSc in Physiotherapy, a M.Phty (Manips.) and an MSc in Applied Biomechanics. Before joining the English Institute of Sport in 2005 to be the clinical lead physiotherapist in London, he worked in the United States in the NFL, NCAA and in the UK in Cricket and Athletics. He worked with the RFUW from 2005 to 2006 under the auspices of EIS, which built on previous work he had done with IRFU and Wasps RFU.

Jenny Page
Department of Sport and Exercise Science, University of Portsmouth

Jenny Page is a Senior Lecturer in Sport and Exercise Psychology at the University of Portsmouth. She is completing a PhD at the University of Liverpool and is a BASES accredited sport psychologist. Jenny has been a consultant for a range of sports and is currently working with performers in gymnastics, rugby league and football.

Chris Shambrook
Director, K2 Performance Systems Ltd Reading

Chris Shambrook has been as a sport psychologist since 1996. He gained his PhD from the University of Brighton, is currently carrying out consultancy work with the highly successful Great Britain Olympic Rowing Team and has been part of the support team at the last three Olympic Games. Past roles have also seen Chris working for the England and Wales Cricket Board, Team Faldo golf, Cambridge and Oxford University Boat Race

Crews, Sunderland Football Club and a number of elite individuals across a variety of sports.

Richard Thelwell
Department of Sport and Exercise Science, University of Portsmouth

Richard Thelwell is a Principal Lecturer in Sport Psychology at the University of Portsmouth. He has a PhD from the University of Southampton and is a BPS chartered and BASES accredited consultant sport psychologist who has worked extensively with a range of sports for over 10 years. Richard is currently actively involved with elite performers in cricket and youth sailing.

Acknowledgements

We would like to thank all those who have contributed to the book, especially the authors who gave up their valuable time to provide excellent case studies of their work. Special thanks also go to Celia Carden at Wiley for her support and guidance through the project.

Brian Hemmings

Many people have helped me develop my skills as a sport psychologist and there are a number of individuals who have been particularly influential. I would like to acknowledge Richard Butler, David Berry, Jan Graydon, Ian Maynard, Jonathan Katz and Hugh Mantle. Special mention also to my co-editor Tim Holder, for his professional expertise over the years and his personal friendship.

Tim Holder

Throughout my career to date, a large number of academics, coaches and applied practitioners have influenced and enhanced my development and helped to keep me 'on my toes' in applied sport psychology. In particular I would like to acknowledge Dave Collins, Keith Davids, Marcus Smith, Carolyn Carr, Tony Morris, Terry McMorris, Iain Greenlees, Ian Maynard, Jan Graydon, Jane Lomax, Andy Balsden, Jerry White, Matt Dicks, Richard Thelwell, Neil Weston, Don Parker, Jill Parker, Peter Hirst, Ben Chell, Jenny Page, Stacy Winter, Bernadette Woods, Sarah Cecil, Paul Dancy, Gill McInnes, Jo Batey, Candice Williams, Amanda Wilding, Christine Johnston, Julie York, Nathan Persaud, Katie Richards and Brian Hemmings.

1
Introduction

Brian Hemmings[a,b] and Tim Holder[b]

[a] Sport Psychology Consultant, London, UK
[b] St Mary's University College, London, UK

1.1 The cognitive–behavioural approach

There are many approaches to applied practice in sport psychology. Poczwardowski, Sherman and Ravizza (2004) stated that 'it is the professional philosophy of a consultant that drives the helping process and determines the points of both departure and arrival regarding the client's behaviour change and also guides consultants in virtually every aspect of their applied work' (p. 446). Accordingly, the case studies detailed in this book are informed by a cognitive–behavioural approach. This does not suggest that this is a more efficacious psychological model, but rather it acknowledges that a cognitive–behavioural approach is adopted by many practising sport psychologists.

Briefly, cognitive–behavioural therapy (CBT) is an umbrella label for approaches originally based on cognitive therapy and behaviour therapy and describes interventions that aim to decrease psychological distress and maladaptive behaviours by modifying cognitive processes (Greenberg & Padesky, 1995; Hill, 2001). This model postulates that individuals' perception of their world is subjective and cognitively mediated, and emphasizes the interaction between current situations, cognitions (what we think), emotions (what we feel) and behaviour (what we do). In practice, the sport psychologist using a cognitive–behavioural approach will engage in a collaborative relationship with their athlete/s. Moreover, the focus of the consultancy will be on 'client' difficulties in

Applied Sport Psychology   Edited by Brian Hemmings and Tim Holder
© 2009 John Wiley & Sons, Ltd

the present and historical information will be gathered only in so much as it has a direct link with the present. According to Scott and Dryden (2003), within this approach the emphasis is upon breaking down negative links between cognition, behaviour and emotion, generally using the cognitive and behavioural 'ports of entry'. In other words, emphasis is upon facilitating change via thought processes and behaviours, the assumption being that there is a direct connection between the two.

1.2 Types of assessments and interventions

The case studies included within this book employ assessment, which is essential to begin the consultancy process. A key impact of assessing athletes, which can often be considered an intervention in itself, is the raising of self-awareness in the client. This is incorporated within many of the case studies within this book either explicitly as a targeted area for development, or implicitly within the assessment process. The range of assessment modes available to the applied sport psychologist can be summarized as interviews, questionnaires (or pen and paper assessments) and observations. The relative strengths and weaknesses of each mode of assessment in eliciting accurate and reliable information establishes the need to triangulate findings from different modes to maximize the accuracy of subsequent intervention decisions (see Beckmann & Kellmann, 2003; Vealey & Garner-Holman, 1998; Taylor, 1995). Once these decisions have been made, the applied sport psychologist can then feel confident to intervene with the performer using a range of techniques (or methods) through which they aim to influence psychological skill, well-being and ultimately performance (Anderson *et al.*, 2002).

1.3 Techniques and skills

The content of the case studies reflect a range of different objectives in the applied consultancy situation. In many cases, however, the delivery of interventions is based on the development of psychological skills. An important distinction to bear in mind whilst reading the case studies is that between the objective of developing psychological *skills* and the *techniques* used to achieve this. Vealey (1988) outlined psychological skills of importance within sport, including self-awareness, self-confidence, optimal attention and optimal arousal. The *techniques* adopted within the case studies such as self-talk, goal setting, imagery and profiling are methods through which the sport psychologist can influence the psychological *skills* of the performer. It is important to recognize that applied sport psychology practitioners may use the same technique to influence different psychological skills and that a number of techniques may well be used in combination with the intention of enhancing a single psychological skill.

1.4 Reflective practice

Each of the case studies includes a section relating to reflections on the process from the sport psychologist's perspective. This is an essential element within all applied work, regardless of the focus of the intervention. Reflection is a process by which knowledge can be developed based on professional practice (Durgahee, 1997). The reflective process can help applied sport psychologists to be more confident in their professional practice in the face of uncertainty (Ghaye & Lillyman, 2000). Therefore, reflective practice helps professionals to learn from experience in a systematic manner and understand that the nature of working as an applied sport psychologist involves uncertainty within dynamic circumstances. Being able, through reflection, to embrace that uncertainty enables the practitioner to interpret this as a challenge. A number of authors have identified systematic, often circular, representations of the reflective process to guide practitioners (e.g. Anderson, Knowles & Gilbourne, 2004; Johns, 1994), some of which are specifically used within certain chapters of the book.

The book is separated into three sections. Section one covers five cases of support work with individual athletes. Chris Harwood reports an intensive intervention programme to build self-efficacy in an emerging professional tennis player; Brian Hemmings details the process of assisting an international test cricketer deal with various on and off field distractions; Andrew Lane describes the use of a videotape intervention with a professional boxer preparing for a world championship bout; Caroline Marlow documents the challenging of limiting performance beliefs with a nationally ranked tenpin bowler; and Iain Greenlees reports an intervention programme to build confidence in a youth golfer.

Section two consists of three chapters on team interventions. Chris Shambrook chronicles interventions with Olympic rowers to enhance communication between crew members; Jenny Page describes her experiences as a trainee sport psychologist delivering group educational workshops to age-group rugby league squads over two seasons; and Richard Thelwell details the effects of a goal-setting intervention within professional football.

The final section includes three chapters on aspects of working with support staff. Jonathan Katz reports the task of developing and delivering psychological support services to athletes and support staff at a Paralympic Games; Sarah Cecil documents the role of the psychologist working as part of a multi-disciplinary team with an injured rugby union player, whilst Tim Holder describes a skill acquisition approach to coach education in table tennis.

The chapters included are not fictional stories or hypothetical scenarios. They are real-life people and real-life events, though in many cases they have been anonymized to protect the identities of the athletes involved. Each case study starts with relevant background information and reports the initial assessment, the psychological intervention/s used, and the monitoring and evaluation of those interventions. Each chapter also includes the reflections of the practitioner on the process of support and their effectiveness, and finishes with a summary and five questions for students.

References

Anderson, A.G., Knowles, Z. and Gilbourne, D. (2004) Reflective practice for sport psychologists: concepts, models, practical implications, and thoughts on dissemination. *The Sport Psychologist* **18**, 188–203.

Anderson, A.G., Miles, A., Mahoney, C. *et al.* (2002) Evaluating the effectiveness of applied sport psychology practice: making the case for a case study approach. *The Sport Psychologist* **16**, 432–453.

Beckmann, J. and Kellmann, M. (2003) Procedures and principles of sport psychology assessment. *The Sport Psychologist* **17**, 338–350.

Durgahee, T. (1997) Reflective practice: Nursing ethics through story telling. *Nursing Ethics* **4**, 135–146.

Ghaye, T. and Lillyman, S. (2000) *Reflection: Principles and Practice for Healthcare Professionals.* Quay Books, Salisbury.

Greenberg, D. and Padesky, C. (1995) *Mind Over Mood.* Guildford Press, New York.

Hill, K.L. (2001) *Frameworks for Sport Psychologists.* Human Kinetics, Champaign, IL.

Johns, C. (1994) Guided reflection. In: Palmer, A., Burns, S. and Bulman, C. (eds), *Reflective Practice in Nursing*, pp. 85–99. Blackwell Science, Oxford.

Poczwardowski, A., Sherman, C.P. and Ravizza, K. (2004) Professional philosophy in sport psychology service delivery: building on theory and practice. *The Sport Psychologist* **18**, 445–463.

Scott, M.J. and Dryden, W. (2003) The cognitive–behavioural paradigm. In: Woolfe, R., Dryden, W. and Strawbridge, S. (eds), *Handbook of Counselling Psychology* (2nd edn). Sage, London.

Taylor, J. (1995) A conceptual model for integrating athletes' needs and sport demands in the development of competitive mental preparation strategies. *The Sport Psychologist* **9**, 339–357.

Vealey, R.S. (1988) Future directions in psychological skills training. *The Sport Psychologist* **2**, 318–337.

Vealey, R.S. and Garner-Holman, M. (1998) Applied sport psychology: measurement issues. In: Duda, J.L. (ed.), *Advances in Sport and Exercise Psychology Measurement*, pp. 432–446. Fitness Information Technology, Morgantown, WV.

SECTION A

Working with Individuals

2
Enhancing Self-Efficacy in Professional Tennis: Intensive Work for Life on the Tour

Chris Harwood

Loughborough University, Loughborough, UK

2.1 Introduction/background information

The purpose of this chapter is to provide students and practitioners with a unique insight into an intensive piece of consultancy conducted with a young professional tennis player as he made the transition towards playing Association of Tennis Professionals (ATP) events on the professional circuit. At the time of this particular block of work, the player[1] was 20 years old and ranked 315 in the ATP world rankings. He had turned professional two years earlier and had reached a career ranking high of 278 seven months prior to the work conducted here. A core responsibility for any sport psychology practitioner is to have a well-developed understanding of the sport within which they are working and its multi-dimensional demands (Boutcher & Rotella, 1987; Taylor, 1995). Therefore, to give an initial context to this case study, it is vitally

[1]The author has received permission and consent from the player to present the elements of work and excerpts discussed in this chapter, and the player has reviewed this manuscript personally.

Applied Sport Psychology Edited by Brian Hemmings and Tim Holder
© 2009 John Wiley & Sons, Ltd

important for readers to first appreciate the ranking system of professional tennis and the decisions that players face when entering tournaments of different levels.

The professional tennis tour

In men's professional tennis, a player's best 14 tournament outcomes in a rolling 12 month period constitute the current points total from which a player's world ranking is calculated at any given week of the year. This means that a player is always under pressure to defend the points that he may have earned in that tournament or week 12 months earlier. If he fails to attain the same number of points (e.g. from reaching the final or winning the event), then his 'best 14 points total' will drop and his ranking may drop, depending on the newly calculated points totals of other players around him. The points total is calculated weekly and the resultant ranking is a critical determinant of the level of tournament that a professional player is able to enter.

The highest level tournaments form the ATP Tour and players typically need to be ranked within the top 100 to have any chance of direct entry into these events. These events (usually two to four per week across the globe) naturally offer the greatest ranking points and the most prize money. Below the ATP Tour is the Challenger circuit, which supports a number of weekly events across the globe. Challenger tournaments are typically the staple diet for players ranked between 100 and 300; they offer less prize money and points and act as a 'staging post' for players to make their move towards the ATP Tour. Below the Challenger circuit is the Futures circuit, which serves as the entry-level structure for young professionals typically ranked from 300 to over 1000. These tournaments offer the lowest points and prize money to players.

The development and progression of the young professional player is a matter not only of collecting ranking points but also of competing 'upwards' against higher standard players to become more physically and mentally accustomed to the nature of the game played by more experienced and mature professionals. A position of around 300 in the rankings means making calculated decisions about whether, for example, to enter a Futures event (where one might be a high-seeded player) or a Challenger event where one's ranking may just be high enough for a direct acceptance into the draw without having to play qualifying rounds. The 'risk/reward' issue is clear – the Futures event may be an easier tournament with modest prize money and points for winning, but the Challenger event provides a much tougher draw for greater points and money to progress and make breakthroughs. However, if one loses in the first round, then zero points and little money are gained for that week of work. In this respect, the professional tennis circuit, particularly at the Challenger level, can be both brutal and demoralizing for the self-employed player.

Contextualizing the consultancy

During 2006, this particular player (from here on referred to as Shane) had competed in a combination of Futures and Challenger events, holding his ranking between 280 and 300. However, during the autumn of 2006, he fully committed to the Challenger circuit and lost narrowly in a sequence of early rounds. These 'near misses' led his ranking to drop to 315 in December as he embarked on a three-week physical training and practice block at a training centre in Florida. This block of training served as preparation for a renewed investment in the Challenger circuit between mid-January and March 2007 alongside qualifying rounds for the ATP event in Doha and the Australian Open Grand Slam in early January.

I had worked with Shane consistently for seven years, providing on-court and off-court sport psychology education and support throughout his junior years and during his transition to the senior game. The consultancy work presented here is therefore located in the context of having cultivated a close, collaborative working relationship with Shane and an in-depth understanding of his knowledge and use of psychological strategies. My philosophical approach to working with him reflected a strong commitment to both humanistic and cognitive–behavioural methodologies (see Hill, 2001; Poczwardowski, Sherman & Ravizza, 2004). I believe strongly in the smooth collaboration of these two approaches to player development, particularly given the precise nature and demands of professional tennis. At the micro-level of a single competitive tennis match, optimal psychological performance is largely a matter of refined self-regulation and the ability to control (and indeed skilfully direct) thoughts, emotions and behaviours as one 'performs and recovers' point by point against the opponent. At the macro-level of player growth and development, it is about dealing with the 'hotel room and treadmill' lifestyle of being a young professional and managing your 'unfolding' identity as an individual throughout this process. Humanistic and cognitive–behavioural orientations ultimately work hand-in-hand in helping the 'person and player' to reach their potential.

Online consulting via e-mail and Skype™ was the regular mode of communication and support when the player was competing abroad. During exchanges in autumn 2006, Shane requested more intensive, face-to-face support during this December training block in the United States. A one-week block of support was ultimately organized with the needs analysis process being shaped by both player and psychologist in a collaborative manner.

2.2 Initial needs assessment

An outcome of working with a player over an extensive period of time is that they become more finely attuned to their own psychological support needs. The educational work with Shane from 13 to 16 years of age had focused greatly on process goal setting and self-monitoring, performance review, mental preparation strategies (including imagery use) and self-management routines (Vealey, 2001). An increasing self-awareness of the psychological demands of tennis and the strategies that would help him to skilfully cope with these demands formed the spine of our work. As Shane began the transition onto the senior tour, the foundation skills for self-regulation were already developed and our work focused on consistently refining cognitions and behaviours in response to particular challenging and demanding situations that he had faced (or would face) on the Futures circuit. Combined with his reflective log book of training and matches, observational feedback from myself, 'working' discussions, on-court behavioural conditioning practices and bespoke mental preparation CDs, Shane honed his self-regulation skills and began to more efficiently decipher and articulate any areas that needed attention.

In the weeks prior to the training block, Shane had lost several close matches against higher-ranked players on the Challenger circuit. These matches essentially represented 'new territory' for Shane. Reflective discussions about these matches were conducted with Shane using a highly task-involving and mastery-oriented consulting style (see Harwood, 2005; Harwood & Swain, 2002). This essentially focuses on the development of abilities and places value on probing self-perceptions and experiences of opponents' skill sets. Several formative points emerged from these discussions:

- Shane was close to creating winning positions and felt able to get his opponents 'on the ropes' on several occasions. He believed that he had the technical and physical 'weapons' to win these types of matches, but that they needed strengthening further.

- The opponents could raise their level at the appropriate time a little better than Shane could, and this was something that he had not encountered often before. Cognitive anticipation of an opponent raising their level had caused Shane to engage in more defensive or protective shot selections at times when he actually had a potential hold on the match (e.g. when leading). This occasional defensiveness was perceived to be inefficient and potentially costly.

- Shane felt that he needed to stay committed to his own game plan and point-by-point process, and assert himself over the opponent more at key times (as opposed to allowing the opposite to happen).

- Shane had been attending to his emotion and behaviour management skills in between points using strategic work that we had done during the summer period. Specifically, we had discussed how his 'game face' or basic psychological character on court represented a disciplined, composed, low positive energy state in between points. However, he was able to activate, react or spontaneously shift to a high positive energy state in response to points that he felt could turn momentum in his direction. To organize this system of behaviour further, a number of physical behaviours (e.g. cadence of walk/footwork; jogging to the chair/towel; giving the opponent a look; clenching a fist; smiling at the audience) were practised and integrated into his routines (Loehr, 1990). He noted the benefit of this work in terms of planning and conditioning optimal internal and external responses to the journey and flow of the match. However, he wanted to practise and develop these skills further.

- Shane was at his most vulnerable emotionally to poor line calls and umpiring, particularly on perceived key points. This caused him to dwell for excessive time on the injustice between points, and sometimes to compromise concentration and gained momentum for subsequent points. Speed of recovery from an emotional perspective was critical at this level; he understood that there was no time for dwelling on 'uncontrollables' as the margins that allowed emotional relapses without punishment were much tighter compared with Futures-level opposition.

- Shane had noticed the benefit gained from executing the wide serve to the deuce court and felt he needed greater mental and physical practice to refine the execution of this weapon.

- Shane articulated the fine margins between winning and losing at this level, and that, although he believed he had the potential to be a top 100 player, he needed to break down some barriers to help him unlock that belief.

This information was valuable in conceptualizing what was going on in Shane's 'world' at this time, and offered a fabric structure to the strategies that would comprise the training week.

Conceptualizing player needs

Shane's own decision to invest in a hard three-week physical training block in December prior to resuming Challenger-level events and qualifying rounds for ATP events provided a few key insights into his character.

Firstly, it spoke to his quality of motivation (both intrinsic and identified regulation; Deci & Ryan, 1985) and resilience, to attempt breakthroughs at this level of the tour when the risks to his ranking were high at that time. Twelve months earlier he had performed well in several back-to-back Futures events and would need to defend these points throughout January and February. If he failed to qualify for or win rounds in any of the upcoming ATP and Challenger events, then his ranking would drop towards the 350s and the possibility of gaining direct acceptances into the main draw of Challenger events would be compromised. He would then have to fight his way through the qualifying rounds of Challengers.

Secondly, and based on previous successful training blocks, he had made the association between improved strength and speed/agility work and the perceptions of confidence that this gave him for his on-court performance. In Shane's mind, enhanced perceptions of strength and a 'no stone left unturned' mental and physical work ethic were key determinants of his subsequent adaptive behaviours and performance on the circuit. In this respect, self-efficacy theory (Bandura, 1977, 1997) played a significant role in helping to understand this particular player.

Drawing from self-efficacy theory

The self-efficacy construct is one of the most influential psychological factors believed to influence achievement strivings in sport (Feltz & Lirgg, 2001). Self-efficacy is defined as 'a belief in one's capabilities to organize and execute the courses of action required to produce given attainments' (Bandura, 1997, p. 3). Unpacking this rather dense definition, self-efficacy beliefs are not judgments about one's skills, but rather they reflect judgments of what an individual can accomplish with these skills (Bandura, 1986). These personal judgments are a result of a complex process of self-appraisal and self-persuasion that relies upon the cognitive processing of diverse sources of efficacy information. Bandura (1977, 1986) categorized these sources into past performance accomplishments or mastery experiences, vicarious experience, verbal persuasion and emotional arousal/physiological states. These sources, in descending order of predictive strength, are all essential sources of efficacy information that determine an individual's level of self-efficacy (or efficacy expectations) for valued tasks.

Individuals with a high level of self-efficacy are expected to demonstrate positive cognitive and behavioural patterns including consistently elevated effort, enthusiasm to approach challenging tasks, persistence in adversity, positive goal intentions, reduced worry and internal, responsibility-oriented attributions. These psychological qualities collude in a fashion that has provided consistent support for the relationship between self-efficacy beliefs and performance (see Bandura, 1997; Feltz & Lirgg, 2001 for a review). Typically, athletes exhibiting high self-efficacy work harder, persist in the task longer and achieve at a higher level over and above athletes who doubt their capabilities (Barker & Jones, 2006).

The themes that emerged from my discussions with Shane certainly did not point towards low levels of self-efficacy or a confidence problem. However, relative to the standard of tennis demanded at this level, there were sufficient indicators to suggest that, like any high-performance vehicle, his self-efficacy beliefs needed 'servicing and maintenance'. His cognitions about opponents at this level, emotional responses to setbacks and subsequent behaviours (in the form of shot selection/decision-making) pointed to discrepancies in the beliefs that would be required to optimize his performance within Challenger-level matches. Note that we are talking here about fine margins and shades of improvement in taking 'good performances' towards 'optimal performance' for his current personal standard. Additionally, he referred to the need to find something that would help him to unlock further his belief in making the top 100.

Within the self-efficacy framework, I could also conceptualize the work required on a couple of other specific psychological sub-themes emerging from our dialogue. These related to further developments in his cognitive and emotional management in between points, and his mental practice of the wide serve (along with physical practice). With respect to the former, a tennis player's perceptions of his emotional state (often coined as arousal control/management) often act as a determinant of self-efficacy in that the more a player feels as if they are effectively regulating or controlling their thoughts and feelings in a competitive situation, the more efficacious they are likely to feel in the execution of the relevant achievement task. With regard to the latter, mental practice (i.e. imaginal rehearsal) of successful wide serves acts as a form cognitive self-modelling (Bandura, 1997; Maddux, 1995). In conjunction with physical practice, these strategies are believed to impact on efficacy beliefs through the sense of repeated accomplishments and task mastery.

Pre-intervention assessment of self-efficacy

As an applied researcher and practitioner, I tend to restrict my use of construct-related, quantitative instruments (for assessment purposes) to those that I am absolutely confident are ecologically valid, context-sensitive and 'fit for purpose'. In professional practice, the self-report information gained from a client's completed questionnaire needs to be specific, contextual and practically meaningful with respect to subsequent work (see Harwood, 2002). Having gained detailed qualitative or observational information from which to interpret athlete needs, it is often useful to ask oneself the question: 'How would a questionnaire now serve to enhance the consultation process?' Normally, the only purpose that it serves relates to some form of 'pre/post', 'then/now' awareness, feedback and monitoring. This is a valid use of such a tool if the process is believed to be insightful for the practitioner and/or client, and only if the tool assesses precisely what you want it to assess. In terms of assessing self-efficacy, its micro-analytic approach requires a detailed assessment of the level and strength of self-efficacy beliefs. Level of self-efficacy relates to beliefs about the level, depth or magnitude of possible task performance, whereas strength relates to the certainty that one has about attaining a particular level or magnitude of performance (Bandura, 1997). Owing to the situation- or sport-specific nature of self-efficacy, bespoke assessments of self-efficacy are typically constructed for use with an athlete.

Table 2.1   Pre-intervention self-efficacy assessment

At the present moment in time, to what degree do you believe that your physical, technical, tactical and mental skills will combine to help you to breakthrough towards the top 200?

Circle the % of certainty that you have next to each ranking level.

Ranking level   **Percentage of certainty (%)**

(Units	1	2	3	4	5	6	7	8	9	**10**	11)	
Inside the top:												
275	0	10	20	30	40	50	60	70	80	**90%**	100	$10 \times 275 = 2750$
250	0	10	20	30	40	50	60	70	**80%**	90	100	$9 \times 250 = 2250$
225	0	10	20	30	40	50	60	**70%**	80	90	100	$8 \times 225 = 1800$
200	0	10	20	30	40	50	60	**70%**	80	90	100	$8 \times 200 = 1600$

Total efficacy score (pre-intervention) = 8400 (2750 + 2250 + 1800 + 1600)

Maximum possible efficacy score = 10450

Pre-intervention efficacy index = 8400/10450 × 100 = 80.3%

With Shane, whilst I was interested in confirming a need for enhanced efficacy expectations, I was as equally interested in using a pre and post assessment of self-efficacy to help me judge the effectiveness of the intervention. On arrival in Florida, I took him through the self-efficacy scale illustrated in Table 2.1.

In this instance, it was important to take a more proximal focus on his beliefs in those skills with respect to breaking into the top 200, and no further. Although Shane's long-term focus was breaking into the top 100, the top 200 had been previously established by the player and coaching/support team as a realistic 12–18 month goal. Therefore, focusing on more distal, less realistic mid-term outcomes such as the top 100 would have been motivationally inappropriate and poor consulting from an attentional perspective. As depicted in the table, Shane's responses to the four ranking levels in terms of strength of belief revealed an 80.3% efficacy index on the day prior to the intervention. This would be compared with the post intervention responses following day 6 of the intervention.

2.3 From identified needs into intervention objectives

Translating the information gathered during this autumn period, intervention work with Shane during the training block focused on one key objective:

- To facilitate an increase in self-efficacy beliefs pertaining to those performance attributes required for ranking breakthroughs on the Challenger circuit.

In order to enhance these efficacy expectations for the 2007 circuit, the intervention would need to target efficacy beliefs related to different sub-components or domains of performance. For this particular period, these included:

- physical strength and conditioning attributes;
- within-match self-regulation skills;
- 'positive' shot selection/decision-making (when leading, under pressure);
- wide first and second service executions.

Consulting strategies would therefore need to be designed that would stimulate increases in these domain-specific efficacy expectations during the physical training and practice block.

2.4 Interventions and monitoring

Bandura's (1997) model of the four sources of efficacy information present the creative practitioner with the potential for a multi-dimensional intervention programme that allows for the different sources to be stimulated or accessed simultaneously. In my model of practice, the sources work together on the player in a coordinated fashion much like a Formula One pit-stop crew when the driver enters the pit lane. Psychological strategies stimulate the sources that work in concert with each other to specifically influence efficacy expectations.

In the forthcoming sub-sections, I will detail the different psychological techniques and methods that comprised the overall intervention and show how they relate to a particular source of efficacy information. However, this will follow an appreciation of the logistics and structure of the week.

Intervention logistics

Unlike typical sport psychology consulting processes, the psychological support work with Shane took place over an intensive week-long period. Traditional clinical boundaries also gave away to some of the realities of field consultations and financial constraints in that both player and consultant stayed in the same apartment, spending the majority of the time on court or in the gym or physical training areas, as well as eating meals together and working through the evenings on the day's achievements. Such an intense seven-day working partnership was facilitated by the existing professional relationship that had been developed over the previous seven years.

I arrived at the Saddlebrook Training Center in Tampa, Florida on a Saturday evening for the middle week of a three-week training block. Shane had trained for one week with a strength and conditioning coach provided by the Center and practised twice daily with a cluster of other tennis professionals using the Center. A similar structure to the week would follow with a gym or on-court fitness session followed by a tennis practice in the morning (9.00 a.m. to 12.00 p.m.) and a further gym or physical training session in the afternoon, followed by an additional hour of tennis practice and recovery (2.30 to 5.00 p.m.). Psychological techniques and process goals (see intervention strategies) were purposely integrated within and between these sessions, often supporting the natural work of the conditioning coach, as well as ensuring that every tennis practice contained agreed psychological outcomes. A portion of time each evening was devoted to debriefing the day (see Accomplishment Debrief below), mental rehearsal and process planning for the following day. The structure, process and content of the entire week are summarized in Table 2.2. The following sub-sections describe the psychological strategies that were developed and consistently integrated during the week.

Intervention strategies

The reflective interview

While Sunday was a rest day from physical training, it was not a rest day from psychological work. In order for Shane to recall and capture some of his past performance accomplishments and strengths over the previous 12 months, a reflective interview was conducted with him. This placed Shane as the 'authority' on his transition into the senior game and allowed him to reflect on some of the following topic areas:

- best performances and memories during the year;
- areas within his on-court game that had improved;
- differences between levels of the game that he was beginning to master;
- lessons that he had learned about tennis at this level;
- challenges of the circuit lifestyle that he was coping with;
- new strategies that he felt were potentially effective;
- advice he would give to younger professionals;
- reflections of the previous training week;
- his mission and goals for the weeks to come.

Whilst not intending to be a motivational interview (Miller & Rollnick, 2002), the session was designed to bring his strengths and capabilities closer to the goals that he

Table 2.2 Overview of weekly training schedule and support: Saddlebrook, Florida, December 2006

Sunday (day 1)	Monday (day 2)	Tuesday (day 3)	Wednesday (day 4)	Thursday (day 5)	Friday (day 6)	Saturday (day 7)
Morning	Morning	Morning	Morning	Morning	Morning	Morning/afternoon
Completion of self-efficacy assessment	Pre-session imagery (20 min)	Pre-session imagery (20 min)	Priming using Alphabet Motivator	Pre-session imagery (20 min)	Priming using Alphabet Motivator	Day off training
The reflective interview	On court SAQ session	Physical training: Weights/core	On court SAQ session	Physical training: Weights/core	Physical training: Gym and SAQ	Discussion and reassessment of self-efficacy
Finalization and agreement of mission and goals for the week	Tennis practice with pros	Tennis practice with pros	Tennis practice with pros	Tennis practice with pros	Tennis practice with pros	Afternoon: Golf (matchplay): lost 2 and 1!
	Strategies: • Alphabet Motivator • mastery video • social support	Strategies: • Alphabet Motivator • mastery video • social support	Strategies: • Alphabet Motivator • mastery video • social support	Strategies: • Alphabet Motivator • mastery video • social support	Strategies: • Alphabet Motivator • mastery video • social support	
Afternoon/evening	Afternoon	Afternoon	Afternoon	Afternoon	Afternoon	Evening
Alphabet Motivator (development of A–F)	Physical training: Weights/core	Physical training: Weights and SAQ	Physical training: SAQ and interval training	No physical training Wide serving practice	Physical training: Weights/core	Flight (Tampa to Heathrow)
Golfs (Matchplay): lost 3 and 2!	Tennis practice: Self-regulation training	Tennis practice: Wide serve work; Self-regulation training	Tennis practice with pros	Conditioned practice sets (U-14)	Tennis practice with pros	Coach arrives to reinforce work done for following week
Evening	Evening	Evening	Evening	Evening	Evening	
Refinement of imagery script in readiness for day 2	Accomplishment debriefing	Accomplishment debriefing	Accomplishment debriefing	Accomplishment debriefing	Accomplishment debriefing	
	Alphabet Motivator (development of G–L)	Alphabet Motivator (development of M–R)	Alphabet Motivator (development of S–Z)	Rehearsal of full Alphabet Motivator		
	Imagery script work					

wanted to achieve. The conversation with Shane served to stimulate recall of positive assets, memories and experiences located in the context of the professional circuit. The goal of this exercise was to allow recent *past* mastery experiences, a key efficacy source, to rise to the conscious surface, and for these to be used in Shane's *current* expression of his mission and goals for the training block. Note that this technique was aided by Shane's verbal articulation level and provided a rich source of information that I was able to directly employ within some of the imagery work described later.

The alphabet motivator

In previous work I had introduced Shane to a list of inspirational and meaningful quotes or statements which started with each letter of the alphabet (e.g. *Ask* yourself 'what can I do better?' on a regular basis; *Communicate* your thoughts and ideas without fear; *Excellence* leaves a trace . . . so leave a lasting memory for yourself and others). Shane had found some of these statements cognitively stimulating for both practice and competition situations, but they were not his own agreed statements. Therefore, aided in part by further information from the *reflective interview*, we developed an alphabet list of short statements which comprised cognitive affirmations, self-instructions and personally symbolic reminders. These cue phrases were designed to facilitate behavioural intensity (and hence, performance accomplishments), attentional and emotional control for both physical training sessions and on-court self-management practices between points during the week. This tool also served as a key vehicle for enhancing efficacy through the mechanism of 'verbal persuasion' even though it was self-persuasion in this case.

On day 1, we developed *A to F* and only the cue phrases from these letters could be covertly or overtly called upon during the following day's training and practice. We progressively continued this project, developing six or seven more letters per day through the next four days. The aim by day 6 was for Shane to be able to select and use particular cue phrases to match the situation that he was facing [e.g. a final set of heavy weights in the gym; an opportunity to beat his last performance on an SAQ (Speed, Agility, Quickness) drill; serving at 5–4, 0–15 having just broken serve in a practice set; receiving serve when set point down]. The *Alphabet Motivator*, as it was coined, served as a method of verbal persuasion for Shane that aided both confidence and self-regulation in order to optimize achievement on specific tasks. It formed a core project for the week and was constantly reinforced during the time spent together (e.g. the Center's freezing pool was used for 'ice bath recovery' and Shane had to continuously visualize and overtly recite the growing alphabet list during each two-minute cold phase!). Selected statements from the Alphabet Motivator are displayed in Table 2.3.

Table 2.3 Selected cognitive and emotive statements from Shane's Alphabet Motivator

A ttitude of Iron	
	C onsistently Uncomfortable
H eavyweight Everything	
	J ust Fxxxxxx Nails
Q uestions Answered	
	R elentless Resilience
T otally Ruthless	
	U ncompromising
W orld Class	

Mastery imagery

Guided mental rehearsal exercises were conducted with Shane on a daily basis prior to the physical training and on-court practices. Response propositions (Lang, 1979) associated with Motivational General Mastery Imagery (MGM; Martin, Moritz & Hall, 1999) were integrated into each script as these have been strongly associated with confidence enhancement in athletes. Each script essentially performed a mental readiness function, focusing on active recall of Shane's strengths and rehearsing the successful execution of technical, tactical, physical and psychological responses that would be demanded of Shane in the upcoming session. The imagery exercises utilized mastery experiences and verbal persuasion as sources of efficacy information. The Alphabet Motivator was incorporated by asking Shane to visualize scenarios where he would use each statement to trigger successful achievements in the session ahead.

Additionally, vicarious experience was employed as a source of efficacy by immersing Shane in a *reverse role model possession* scenario. Specifically, this involved progressively stimulating Shane with confidence and control-related images associated with the technical, physical and psychological aspects of their game, and then prompting Shane to select and imagine a top 150-ranked role model. Shane was then asked to imagine sliding himself into the body of that player and contemplate the additional skills that this type of 'possession' would bring to the role model. Subsequently, he was asked to reverse the possession process and consider what (if any) differences the role model player would add to his own skill set. This cognitive exercise was designed to close (if not shatter) any perceived gap that existed between the level which Shane was currently at and the level he was fully capable of reaching.

Three of the 20 minute guided imagery sessions took place in the apartment prior to either a morning or afternoon training session. Following the first imagery session on day 2, each imagery script evolved on a day-to-day basis and was responsive to the previous day's events and achievements. In addition to these structured sessions, guided imagery cues were also used throughout the 10 minute walk to the training venue in order to fully prime Shane towards an ideal psychological state. This 'conversational' work involved exchanging information that was high in cognitive and motivational content (see Hall, 2001) as we casually walked together. Figure 2.1 presents selective segments of the imagery script employed on day 4, which incorporated the second segment of the Alphabet Motivator and the reverse role model exercise.

On-court self-regulation training

Shane's self-management skills during the 'dead time' periods within on-court practice were given a high priority. Every on-court tennis session emphasized the practice of agreed psychological processes and routines. For example, breathing techniques and agreed body language responses were observed whenever there was a break in play. This included responses during warm-up and drilling segments whenever a rally between the two players broke down (i.e. before the start of the next rally), as well as when Shane responded after competitive points in practice tie-breaks, sets or matches. In this respect, whenever 'dead time' appeared in a session, it was viewed as an opportunity to make the dead time 'live' in a positive, functional way for Shane. These moments of time represented opportunities to self-regulate and prepare appropriately for the next point.

A core focus of this on-court work was the integration of cognitive elements to Shane's practice of routines during the 'dead time'. Practising the experience of functional thoughts and affirmations, leading into healthy emotions and assertive body language, sought to improve Shane's perceived control of his emotional state – the fourth source of self-efficacy. The Alphabet Motivator acted as a cognitive source for Shane in that he was asked to practise thinking and using the Alphabet statement that best suited the attitude and preparation required for the next point or effort. This was designed for Shane to experiment with finding and using the cognitive trigger or prime that would help him to emotionally process the last point and tactically prepare for the next. For example, in response to winning a point in which he exerted his strengths over the opponent, he may have selected the statement *Consistently Uncomfortable* as a trigger indicative of constantly pushing the opponent out of their comfort zone and giving them as much trouble and suffering as possible.

As an ex-national standard tennis player, I often work as the player's opponent and adopt a situational mental coaching role where we play out different competitive scenarios against each other. This provides an opportunity to use the court as a field teaching laboratory where routines can be conditioned in between points, observations can be made of player responses and match situations can be 'frozen' to allow for player–consultant discussions of thoughts and feelings at that moment.

Lie back onto the couch and find a position where you begin to feel really, really comfortable.

I want you to close your eyes and focus all of your attention on your breathing... locking yourself into that normal, relaxed breathing rhythm that you have mastered when opening your imagination.

In your own time, begin to take yourself towards the court and the gym where all of today's achievements are going to happen... Imagine walking along the same path through the clusters of green palm trees towards the courts. It's another beautiful day and the sun is warming you up ...

In your mind as you walk along, you are focusing on the next phase of the Alphabet Motivator... G is for *giant steps*... H means *heavyweight everything*... As I go through each of these, begin to think about the positive actions associated with these words in your tennis and the physical training that you've been doing ...

... You've now arrived at the court for the 1st fitness session of the day and again I want you to tune into your professional approach... the fact that you've turned up every single day, mentally and physically to deliver... In every single fitness session, you've pushed yourself through the pain barrier and yesterday you improved on both training technique and your personal best times.

I want you to place yourself in the context of the fitness drills today... and in this next minute I want you to pick one typical drill and any letter or letters you choose (from G to L) for that particular drill... Picture yourself and feel yourself accomplishing the drill... mentally and physically applying yourself using your chosen alphabet statement... using it to focus all of your resources.

I want you to feel the massive sense of achievement from this session... remembering the statement from Mike (the trainer) that you train like a top 100 player... and achieve the same levels of the top 100 players that he has worked with.

You now feel a complete readiness to go onto court, and I want you to imagine yourself on court alive with energy during the warm up. Lock yourself into that high quality of shotmaking from the back of the court... adding your feel of the ball to every quality shot... where every quality stroke has that strength and trust behind it... trust that becomes stronger and stronger and stronger, as you make the stroke... just the way you want to... just how you imagined it ...

Figure 2.1 Example script segments from day 2 imagery.

...   I want you think about yourself being on court as a player who possesses physical strengths in all areas...to a supreme level...and what it feels like to be that kind of player on that court and the sense of control it gives you. A player with a crisp routine in between points...solid...understated...emotionally controlled...with a body language that consistently shows the character behind the player...

And today I want you to imagine that your reactions to everything show a powerful acceptance of some of the inevitable adversities and challenges that you will successfully overcome...Accept the great play as well as the mistakes and the fallabilities of yourself and others. What before might have been a petulant reaction to a mistake now becomes a seamless progression to the next opportunity...the line of trust, of strength, of quality is not broken...because your reaction is intelligent...it is bulletproof.

With all these qualities in mind, I want you to try and imagine yourself playing against a top 150 player that you know. I want you to consider the skills that this player has and then to observe the kind of player that he would become if you slid all of your qualities into his body...how much extra mental and physical weaponry would he have ...?...Now, imagine him sliding into your body with the qualities he has. How little or extra would he add? I want you to notice how you add extra qualities to his game...and how many of the qualities of your opponent are cancelled out by your own strengths.

As the time for your session approaches, I want you to focus on the G to L one more time, playing through those game scenarios where you will use each letter to become the most assertive, intelligent, confident and controlled player required for that specific moment...for that specific situation.

Figure 2.1 (continued)

Two on-court sessions during the afternoons of days 2 and 3 were conducted directly with Shane in this manner. The focus of these sessions lay in the repetitive practice of positive cognitive and behavioural responses to different forms of adversity, as well as the practice of aggressive, decisive shot selections 'when the ball was there to be hit'. For example, in both sessions we played a practice set where Shane started 0–30 down every game of the set against me. This handicap condition would actually create a very even match between the two of us and require a high level of attentional and emotional control on the part for Shane if he was to challenge for the set. In this type of conditioned practice, I also integrated a maximum of two 'tight' line calls where I called balls 'out' that were perhaps 'in'. I usually made these decisions on points which might have earned Shane the game or put him in further trouble (e.g. break point down; 30–40). Throughout the set, I would observe his responses and decisions as an opponent, choosing occasionally to come out of my 'opposition role' in order to interject or freeze the game for discussion and review in my role as on-court mental coach.

This role shifting had to be skilfully managed. If I sensed a critical moment or period in the practice set that called for particular cognitive and behavioural responses, I took one of two options. I would either continue in my role as the opponent, let the flow of our competitive match play out and observe how Shane responded. Alternatively, I would interrupt the game and initiate an exchange with Shane about his perception of, and feel for, the situation in order for us to explore the optimal cognitions and behaviours for the specific period ahead. With the former option, I would review his behaviour and exchange ideas at the end of the period, having observed how he coped with the situation.

Both of these options served to enhance match awareness and placed Shane's ability to think assertively at the centre of his emotional regulation in between points as the journey of the match unfolded. These psychological process goals also formed other on-court practice sessions when Shane trained or competed against professional opposition. In these practices, I played a purely observational, support and ultimately debrief role (see below).

Pre-play serving practice

Time during the afternoon on-court session was devoted to extended serving practice with a specific focus on the wide serve into the Deuce court. Beyond physical practice, however, I encouraged Shane to engage in 'pre-serve mental practice' where he mentally rehearsed the feel, form and successful placement of the wide serve prior to actual service execution (see Morris, Spittle & Watt, 2005). A distributed practice schedule was typically employed with Shane occasionally mixing in normal serving patterns (i.e. T-serves down the centre line, body serves) and practising in the Ad court before moving back to the Deuce court for a further block of work. Practice points would also be played against Shane where serving an ace out wide was rewarded with two points. Pre-serve mental practice was encouraged on each occasion here.

Conditioned practice sets

An opportunity on day 5 arose with the Under 14 LTA British national squad, who had arrived at the Center during the week. Specifically, following serving practice, we organized for two of the players to compete in a live practice set with Shane (i.e. one after the other). The 0–30 handicap condition was once again employed in favour of the youngster and the audience of players and coaches added that extra layer of competitive reality. Shane was encouraged to stay process-focused and to use the opportunity as a competitive practice to implement the work that he had been focusing on (e.g. Alphabet Motivator; wide serve patterns). A tactical process goal was also agreed in terms of accomplishing that aggressive, decisive shot selection when the 'ball was there to be hit'.

Shane regulated himself exceptionally well under these conditions, maintaining elevated levels of concentration in balancing the focus on aggressive shot-making that would unlock the point, with the need to minimize errors given the opponent's game score advantage. Indeed, the Under 14 players were excellent 'cannon fodder' in that their weight of shot created natural opportunities for Shane to practise aggressive decisions, but they were sufficiently consistent to never allow complacency to develop within him. On two occasions, Shane faced consecutive set points against him, coming back from 0–40, 4–5 down on one occasion to win the match. The self-management demonstrated in these situations represented only a small portion of the accomplishments during these sets. The key objective here was cementing the association between his ability to manage his emotional state and resultant performance outcomes. This exercise therefore aimed at stimulating two sources of efficacy simultaneously, namely performance accomplishments and the positive control over emotional states linked to those accomplishments. These achievements were also captured on video. This is a strategy to which we now turn.

Mastery videos

Videotaping sessions throughout the training week was a fundamental process goal for myself as a consultant. Mastery experiences (i.e. performance accomplishments) could be captured and reviewed on an on-going basis, and used by Shane to relive everything (i.e. thoughts, feelings and behaviours) associated with successful task executions and session process goals. His reflective interview, imagery sessions, strength work in the gym, SAQ work outside, on-court practices and the conditioned sets were all videotaped to form an archive of footage that was selectively used during the evening debriefs and as files to view at Shane's convenience. Specific attention was given to reinforcing both process- and outcome-oriented accomplishments in the conditioned sets and certain SAQ efforts where measurable improvements in speed and agility were recorded. Where appropriate, I included an audio commentary alongside encouraging remarks so that a verbal record of my own thoughts and perspectives followed Shane's activities, behaviours and achievements.

Persuasive social support

Verbal guidance, encouragement, priming and commentary (via the aforementioned strategies) represented diverse forms of persuasive social support to influence Shane's efficacy beliefs. Bandura (1997), notes that verbal persuasion exerts its strongest impact on self-efficacy when the 'persuader' is viewed as credible, trustworthy and genuine by the receiver of the information. The working or therapeutic alliance between consultant and player is therefore of significant importance in building up a trusting relationship where spoken words communicated to the player are believed by the player. In addition, as noted earlier, Shane acted as his own persuader through the application of cognitive and motivational self-talk (see Hardy, Gammage & Hall, 2001) associated with the Alphabet Motivator.

Monitoring via accomplishment debriefing

At the end of every training day, a 30 minute period was intentionally devoted to reviewing the levels of mental and physical effort applied during each training session, and exchanging perceptions of any relevant task accomplishments. Shane rated his levels of physical effort, concentration, resilience and persistence associated with the strength and conditioning work and on-court practice tasks on a self-referenced scale of 0–100% of personal excellence. Whilst these ratings acted as one method of monitoring Shane's investment in the intervention, their primary role was to help Shane to focus upon, process and internalize his high levels of mental and physical output on a daily basis. Put simply, these psychological attributes were consistently reported as high (e.g. 85–100%) and scoring them essentially served as a 'process-oriented checklist' to confirm his self-referenced achievements (i.e. mastery experiences) and discuss any discrepancies for that day. The accomplishment debriefing also involved exchanging positive observations on specific behaviours or events in sessions, including use of the (mastery) video as a self-modelling tool for review and reinforcement. All of these exchanges inevitably facilitated any action planning for the following day on areas that needed further attention.

The full schedule of training activities and psychological strategies over the course of the week is illustrated in Table 2.3.

2.5 Evaluation of interventions

The day of my departure (day 7) was a day off from training and an opportunity to review the week, plan for the following (and final training) week (supported by the arrival of his coach), reassess self-efficacy and engage in a competitive leisure activity. The overall evaluation of the intervention included both immediate (i.e. proximal) perceptions of Shane (including his efficacy reassessment), as well as more distal, longitudinal behaviours and outcomes such as Shane's continued use and application of strategies, his perceptions of tournament performances and, most objectively, improvements in ranking.

Table 2.4 Pre and post intervention self-efficacy beliefs

Ranking level	Pre-intervention efficacy strength (%)	Post-intervention efficacy strength (%)	Efficacy scores (pre)	Efficacy scores (post)
275	90	100	2750	3025
250	80	100	2250	2750
225	70	90	1800	2250
200	70	85	1600	1900
Total efficacy score (**and efficacy index**) =			8400 (**80.3%**)	9925 (**95.0%**)

Self-efficacy reassessment

The self-efficacy reassessment revealed an enhanced self-efficacy index of 95% compared with 80.3% at the start of the week (see Table 2.4). This index represents a relative measure of the belief that Shane possesses about his multiple skills taking him from his current ranking towards to the top 200 in the world. The reader will notice that his percentage confidence ratings (i.e. efficacy strength) at progressive ranking levels are higher post intervention. He now believed 100% that his abilities would take him into the top 250, and in practical terms this could be viewed as a significant psychological improvement despite his being over 80% confident of reaching this level at the start. Moreover, at the top 225 and 200 levels, he was 20 and 15% respectively more efficacious about these stages compared with day 1.

Player perceptions and E-mail validations

At the end of the training block, Shane noted the value of integrating intensive mental and physical training activity as well as the intensity and closeness of the working relationship. During the following two months of tournaments, Shane maintained regular e-mail contact including updates that validated the effectiveness of the training block. A selection of these are presented below:

> I've been looking at the DVDs in the last few days. I enjoyed watching the footage and was surprised how motivating I found some of it. There is just a constant stream of positive reactions and body language when I am playing. The footage of me doing weights in the gym is extremely motivating. It's great to have when I want to watch it. XX [touring coach] is going to video me much more now, even if the camera is only on me. I really enjoy watching myself play and the more footage I have, the better.

> I beat XX today 4 and 3. Solid performance. I played to make him feel CONSISTENTLY UNCOMFORTABLE . . . the Alphabet Motivator is coming in handy!

> Hi Chris, great performance in 1st round today. Came out of the blocks like Maximus from Gladiator and hit my opponent with relentless mental resilience and huge groundstrokes.

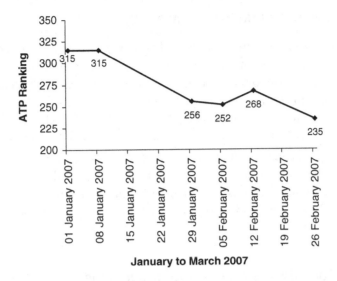

Figure 2.2   ATP ranking history (January–March 2007).

I just gave him too much heat! I have been working on winning the collision with the ball and I need to be dominating the ball and demanding that it does what I want. This comes from noise and the noise comes from a fired up, energetic and motivated mindset. Before my match yesterday I did some imagery on this – I think its going to be an effective tool in this particular area.

These examples reinforced the value of cognitive techniques such as the Alphabet Motivator and imagery (perhaps working hand-in-hand) as well as the self-modelling properties of mastery videos.

Performance and ranking improvements

Over the course of six tournaments in January and February 2007, Shane recorded 12 wins to seven losses at Challenger/ATP qualifying level including wins against players ranked 116, 180, 192, 195 and 220, and two players in the 250s. By reaching both a semi-final and a final in two of these Challenger events, he not only defended but increased his best 14 tournament points total during this period. His ATP ranking moved from 315 to a career high of 235 (see Figure 2.2) at the end of February 2007.

2.6 Evaluation of consultant effectiveness/reflective practice

The experience of conducting an intervention under such circumstances offered me a number of valuable reflections for future practice.

Structured work vs reactive counselling

After working over an extensive number of years with a player, it can be easy for an education-focused practitioner to evolve into the role of merely being a counsellor, mentor or critical friend to the client at the expense of evidence-based and structured performance enhancement work. It is perhaps true that consultant–client relationships evolve as the client grows and develops through the psycho-educational process to points where the consultant may play out the role of 'wise old man' more often than an 'educational consultant with a plan'. However, one of the main reflections from this experience is the importance of ongoing needs assessment leading to evidence-based educational consulting. In this case, the development of fresh ideas and strategies within intensive, planned field-work aided the client during one of the key transitions in his career.

Adaptability within a manageable programme

The need to be reactive and responsive was also challenging. Field consultants have to work 'on the hoof' and be adaptable to environment and circumstances. The development of imagery scripts based on new information and events meant working to a next morning deadline. Further, the uncertainty of which professionals would be practising in any given session (and what they wanted to work on) meant having other plans prepared in advance. For example, the day 2 morning practice session included a top 10 ATP player and four other professionals at various levels, some of whom may have wanted to drill and others who sought to play competitive practice sets. Interjecting and integrating into drilling scenarios and 'hand-fed' points is far easier as a consultant compared with interrupting the flow of a competitive practice set for the opponent.

Whilst the actual 'apartment time' for mental training work (e.g. imagery; accomplishment debriefing) amounted to between only 30–60 minutes per day, we both reflected on the fullness of the days. For example, it would have been easy to have suggested more imagery time, but the fact remained that both player and consultant felt the fatigue by the end of the day. It was critical to plan switching off with leisure time into the week.

Team role vs dual role

I was aware of maintaining my role as a sport psychologist without overstepping into the role expected of a coach. Readers might be surprised by my involvement in tactical work and serving practice even though this was related to psychological strategy. However, this was a feature of the multi-disciplinary support team built around Shane that had worked together for the previous five years. Decision-making and serving work had been openly discussed throughout the team with shared ideas about solutions (see Reid, Stewart & Thorne, 2004). The fact that I was a qualified professional coach and player was secondary to the trust that had been built up amongst the full coaching and support team. The need to know one's boundaries and competencies is critical, as is the need to be able to reinforce and assist the work of other practitioners.

Contextual efficiency

The strategies incorporated into the intervention utilized a variety of contexts consistently: morning and evening time in the apartment; walking to practice; supporting and videoing in the gym and on court; and in the ice bath/swimming pool. Whilst I had some switch-off time as a consultant and did a lot of observational work, I also felt a sense of momentum building throughout the week as a result of the ongoing work that was being accomplished in each context. Every context included a strategy designated to facilitate a psychological effect on Shane. However, all of the strategies fuelled by the self-efficacy model began to combine and integrate efficiently as the week progressed. Imagery sessions and persuasive support reinforced the Alphabet Motivator, Shane's use of Alphabet Motivator was captured on the mastery video and the mastery video captured all accomplishments that might be reinforced later through imagery and persuasive support. In this way, there was a degree of contextual efficiency in implementing techniques that would positively influence Shane's thoughts, feelings and behaviours.

2.7 Summary

The purpose of this chapter was to describe an intensive intervention conducted with an elite tennis player who was making a transition through the Challenger circuit of professional tennis. A few key summary points are worth emphasizing. Firstly, the consultant's knowledge of the professional circuit and the idiosyncratic demands of tennis at this level were critical to the precision and content of the intervention. Students are encouraged to appreciate the need for understanding the micro- and macro-demands of a particular sport. Secondly, a consultant's knowledge of theory combined with a thorough needs assessment are vital in conceptualizing and determining the appropriate goals and strategies for the intervention period. Conducted during a predominantly physical training block, the consulting work in this case sought to facilitate an increase in self-efficacy beliefs pertaining to those performance attributes required for ranking breakthroughs on the Challenger circuit. These efficacy beliefs were targeted using strategies that honed in on:

- physical strength and conditioning attributes;
- within-match self-regulation skills;

- 'positive' shot selection/decision-making (when leading, under pressure);
- wide first and second service executions.

A variety of cognitive and behavioural techniques worked on these domain-specific efficacy expectations during the physical training and practice block. Theoretically, performance accomplishments/mastery experiences represent the strongest source of efficacy information (Bandura, 1977) and therefore virtually all of the strategies aimed to reinforce these experiences in one way or another. Reassessment of self-efficacy, as one method used to ascertain the effectiveness of the intervention, revealed increases in Shane's overall beliefs about these skills.

Finally, the long-term working relationship and alliance with Shane (and his coaching team) helped to optimize the intervention on logistical grounds. This leads to my final conclusion that, when it comes to interventions focused on building self-confidence in various athletic skills, it is useful to have a coach and strength and conditioning trainer working in tandem on exactly the same messages. In this respect, long-term multi-disciplinary collaborations should be highly sought after by athletes, coaches and sport scientists.

Questions for students

1 What is self-efficacy and why is it a critical factor in an elite, professional sport such as tennis?

2 When consulting with a professional athlete, how would you go about assessing and monitoring self-efficacy?

3 What technical competencies, knowledge and attributes does the sport psychologist need to possess when working (a) with a professional athlete and (b) in an intensive training camp context?

4 What are the primary techniques used by consultants in enhancing self-efficacy through its associated sources?

5 How would you evaluate the effectiveness of your work with a professional athlete?

References

Bandura, A. (1977) Self-efficacy: toward a unified theory of behavioral change. *Psychological Review* **84**, 191–215.

Bandura, A. (1986) *Social Foundation of Thought and Action: A Social Cognitive Theory*. Prentice-Hall, Englewood Cliffs, NJ.

Bandura, A. (1997) *Self-Efficacy: The Exercise of Control*. Freeman, New York.

Barker, J.B. and Jones, M.V. (2006) Using hypnosis, technique refinement and self-modeling to enhance self-efficacy: a case study in cricket. *The Sport Psychologist* **20**, 94–110.

Boutcher, S. and Rotella, R. (1987) A psychological skills education program for closed skill performance enhancement. *The Sport Psychologist* **1**, 127–137.

Deci, E.L. and Ryan, R.M. (1985) *Intrinsic Motivation and Self-Determination in Human Behavior*. Plenum Press, New York.

Feltz, D.L. and Lirgg, C.D. (2001) Self-efficacy beliefs of athletes, teams, and coaches. In: Singer, R.N., Hausenblas, H.A. and Janelle, C.M. (eds), *Handbook of Sport Psychology* (2nd edn), pp. 340–361. John Wiley & Sons, New York.

Hardy, J., Gammage, K.L. and Hall, C.R. (2001) A descriptive study of athlete self-talk. *The Sport Psychologist* **15**, 306–318.

Hall, C.R. (2001) Imagery in sport and exercise. In: Singer, R.N., Hausenblas, H.A. and Janelle, C.M. (eds), *Handbook of Sport Psychology* (2nd edn), pp. 529–549. John Wiley & Sons, New York.

Harwood, C.G. (2002) Assessing achievement goals in sport: caveats for consultants and a case for contextualization. *Journal of Applied Sport Psychology*, **14**, 380–393.

Harwood, C.G. (2005) Goals: more than just the score. In: Murphy, S. (ed.), *The Sport Psych Handbook*, pp. 19–36. Human Kinetics, Champaign, IL.

Harwood, C.G. and Swain, A.B. (2002) The development and activation of achievement goals in tennis: II. A player, parent, and coach intervention. *The Sport Psychologist* **16**, 111–137.

Hill, K.L. (2001) *Frameworks for Sport Psychologists*. Human Kinetics, Champaign, IL.

Lang, P. J. (1979). A bio-informational theory of emotional imagery. *Psychophysiology* **8**, 862–886.

Loehr, J.E. (1990) Providing sport psychology consulting services to professional tennis players. *The Sport Psychologist* **4**, 400–408.

Maddux, J.E. (1995) Self-efficacy theory: an introduction. In: Maddux, J.E., (ed.), *Self-Efficacy, Adaptation, and Adjustment: Theory, Research, and Application*, pp. 3–33. Plenum Press, New York.

Martin, K., Moritz, S. and Hall, C. (1999) Imagery use in sport: a literature review and applied model. *The Sport Psychologist* **13**, 245–268.

Miller, W.R. and Rollnick, S. (2002) *Motivational Interviewing: Preparing People to Change*. Guilford Press, New York.

Morris, T., Spittle M.and Watt, A. (2005). *Imagery in Sport: A Complete Picture.* Human Kinetics, Champaign, IL.

Poczwardowski, A., Sherman, C. P. and Ravizza, K. (2004) Professional philosophy in sport psychology service delivery: building on theory and practice. *The Sport Psychologist* **18**, 445–463.

Reid, C., Stewart, E. and Thorne, G. (2004) Multidisciplinary health science teams in elite sport: comprehensive servicing or conflict and confusion? *The Sport Psychologist* **18**, 204–217.

Taylor, J. (1995) A conceptual model of the integration of athletic needs and sport demands in the development of competitive mental preparation strategies. *The Sport Psychologist* **9**, 339–357.

Vealey, R.S. (2001) Understanding and enhancing self-confidence in athletes. In: Singer, R.N., Hausenblas, H.A. and Janelle, C.M., (eds), *Handbook of Sport Psychology*, pp. 550–565. John Wiley & Sons, New York.

3

Managing Distractions in Test Cricket

Brian Hemmings

Sport Psychology Consultant and St Mary's University College, London, UK

3.1 Introduction/background information

The support work that will be described in this chapter involves an international test cricketer over a two-year period. The relationship evolved through my psychological support role for the professional English county club he was appearing for as an overseas player. At the time I had already been working with the club in a consultancy capacity for two seasons. The work with the club comprised facilitating pre-season and monthly in-season squad meetings, group work and individual support. The broad philosophy of the support provision for the team was:

- to eliminate misconceptions regarding the role of the sport psychology consultant and to effectively gain entry into the environment and establish credibility, trust and rapport (Andersen, 2000; Hemmings, 1999; Ravizza, 1988);

- to adopt a cognitive–behavioural framework, and to make sport psychology principles as cricket-specific and individual as possible (Bull, 1995, 1997; Bull, Shambrook, James & Brooks, 2005; Müller & Abernethy, 2005; Müller, Abernethy & Farrow, 2006);

Applied Sport Psychology   Edited by Brian Hemmings and Tim Holder
© 2009 John Wiley & Sons, Ltd

- to follow a model of equal expertise (Butler, 1989) whereby it is assumed that all individuals (players, coaches, other support staff) within the sport bring their own experiences and valuable expertise;

- to assess cricketers' psychological strengths and weaknesses and develop, implement and monitor individual mental training programmes;

- to apply interventions at a team level to improve cohesion and performance (Carron, Hausenblas & Eys, 2002);

- to abide by the British Association of Sport and Exercise Sciences and British Psychological Society codes of conduct, and respect client confidentiality at all times.

The underlying philosophy of service provision was based on previously successful consultancy programmes (e.g. Andersen, 2005; Bull, 1995; Halliwell, 1990; Hardy & Parfitt, 1994; May & Veach, 1987), the mental skills training used recognized applied sport psychology models (Boutcher & Rotella, 1987; Thomas, 1990). However, Hardy, Jones and Gould (1996) have indicated that although consultancy work can sometimes operate in accordance with these step-by-step models, the complex nature of consultancy often prevents sequential-type approaches. The particular case study that I focus upon demonstrates the sometimes cyclical and dynamic nature of support work where the consultant can frequently fluctuate between assessment and intervention as new challenges arise for the athlete.

The cricketer (hereafter named Alan) had been playing for the county for four months, and was performing well in English county cricket in the four- and one-day versions of the game. Hence, there did not seem to be any immediate performance issues. Although I had not conducted any previous individual work with Alan, I felt I had already built a sufficient rapport with the player and he was a prominent voice in team meetings that I had led. Alan approached me about some individual issues after he had been re-called to his country's test squad after a three-year absence. We then had two sessions (approximately 50 minutes each time) in which I listened to his issues of concern and what he wanted the purpose of our sessions to be.

3.2 Initial needs assessment

Interviews

There are many forms of assessment available to the sport psychologist (Taylor, 1995; Taylor & Wilson, 2005). The assessment method used reflected my service philosophy and approach (i.e. often time-limited sessions in a consultation room/changing room balcony/practice ground/net practice). Interviewing involves excellent communication skills, including honed listening skills; the hallmarks of sound counselling skills. By most therapeutic, counselling and clinical psychological standards, the initiation of helping relationships in sport psychology, and the delivery of service, is loose (Andersen, 2000). This is not a criticism of sport psychology, but recognition of the places and ways in which sport psychologists operate, which sometimes present greater variety than the more traditional counselling settings and thus a challenging environment for effective listening (e.g. the use of paraphrasing, summarizing and reflecting skills; Dryden, 2006).

Through historical interviewing (e.g. discussing past playing experiences and performances, career statistics and records) and probing current thoughts (May, 2001; McMullin, 2000), Alan described that whilst he had broken a number of domestic batting records at a young age, his early test career was poor (in terms of runs scored) and he had been dropped twice. He talked of the media pressure in his own country and the opinions of commentators that he would not succeed at test level. He also spoke of what he called the 'scars' left by his earlier experiences and how he felt troubled by the thought of re-entering the test arena to face the same experiences which he felt may affect him *on and off* the field. Through active listening (Palmer & Dryden, 1995) Alan was able to disclose a variety of concerns. These are listed below.

International cricketer's test performance concerns

- 'Scars' of previous test cricket experiences.
- Commentators' opinions on his batting and his ability/suitability for test cricket.
- Selectors' preferences.
- Type of pitches he had to perform on.
- Playing under a captain he felt did not like him.
- Having to 'prove' himself.
- Coach and senior players' 'selfish' views within the team.
- Poor team ethos.
- Lack of support of batting partners at the wicket.
- Media criticisms of him.
- Felt he could not trust other players.
- Attention 'wandering' at the crease.

Alan's first experiences back with the test squad were going to be touring New Zealand and Sri Lanka. There is much debate in professional cricket about the great demands placed on players, both physically and mentally, by the schedule of international test-match tours. The touring environment also held some additional concerns for Alan and through discussion he identified the issues listed below.

Alan's 'touring' concerns

- Time away from home/wife/family.

- Dealing with heat/humidity in Sri Lanka.

- Spending lots of time in hotels/potential boredom.

- Too much time to think about cricket.

- Sharing rooms/lack of personal space and time for long periods.

- Potential for poor practice facilities at some venues.

- Intense schedule and yet possibly not an opportunity in tests, lots of preparation for little playing time.

- Spending lots of time with people I do not necessarily like.

To summarize my assessment of Alan, he was in confident form and displayed many attributes which the literature would suggest are hallmarks of 'mentally tough' cricketers (Bull *et al.*, 2005). However, in his own words he wanted to 'chat through some of the scars' of his previous test cricket experiences and devise new ways to approach his recall to the test arena, on and off the field. My assessment at this point was that Alan was faced with a variety of potential distractions in his performance environment that he needed assistance to acknowledge and deal with more effectively.

3.3 Interventions and monitoring

The ability to control thought processes, to concentrate on a task, is widely recognized as important for effective sports performance (Nideffer, 1999). Performance enhancement interventions used by applied sport psychologists often involve assisting athletes to identify task-relevant aspects of their performance, and also to develop the ability to recognize and deal with distractions from the task. Nelson, Duncan and Kiecker (1993) suggested that a distraction occurs when competing stimuli interfere with or divert attention from the original focus of attention.

Moran (1996) identified that distractions for sport performers essentially fall into two categories: internal (self-generated mental processes) and external (environmental). Internal distractions that have previously been highlighted in the sport psychology literature are thinking about past/future events, a lack of motivation for the task, fatigue, and anxiety. Equally, external distractions have been noted such as noise, crowd size/movement, replay screens, acts of gamesmanship (e.g. 'sledging' in cricket), weather conditions and performing in unfamiliar environments (Moran, 1996; Nideffer, 1999).

Sport psychology literature has provided examples of interventions for applied practitioners to use to help athletes identify and overcome distractions in the sport environment. For example, the use of competition and focus plans (Orlick, 1990; Weinberg & Williams, 1999) and the development of 'so if' (Butler, 1996) and 'do this' (Gordon, 1992) strategies has been suggested. The latter also include cricket-specific strategies.

Focus plans

The first intervention with Alan involved the development of focus plans which incorporated a combination of psychological techniques including motivational self-talk (Theorodakis, Weinberg, Natis, Douma, & Kazakas, 2000), various types of imagery use (e.g. cognitive–general, Munroe, Giacobbi, Hall & Weinberg, 2000) and goal-setting (e.g. Weinberg, Burton, Yukelson & Weigand, 2000). Additionally, Ravizza (1999) has argued that the underlying basis of psychological interventions for performance enhancement in sport involves teaching the athlete the importance of awareness. Indeed, awareness, according to Ravizza, is the first step toward gaining control of a situation. In the context of this case study, raising awareness was seen to be fundamental (see later reflections) in helping Alan deal with internal and external distractions. Moreover, developing self-awareness can also act as an intervention in itself (Ravizza, 1999), and may remove the need for other techniques to be employed by the sport psychologist.

A model for explaining the stages of raising self-awareness (adapted from Robinson's Four Stages of Learning, see McMahon, 2001) that I have used to underpin my work with athletes can be seen below. The essence of this model is that when learning something new (in this case psychological skills), an individual will go through a set sequence of skill acquisition from unconscious incompetence ('don't know it and can't do it') to unconscious competence ('doing things well without thinking about them'). The text that follows the model explains the sequence of raising self-awareness in relation to Alan:

Unconscious Incompetence
↓
Conscious Incompetence
↓
Conscious Competence
↓
Unconscious Competence

In the context of Alan dealing with his distractions, it seemed to be that at first there were some unidentified performance problems causing him concern about his forthcoming re-entry into test cricket (unconscious incompetence). Through listening/interviewing it became clearer to Alan that he was experiencing some internal and external distractions (conscious incompetence) which could interfere with his performance on the cricket field. The realization of this issue enabled Alan to grasp the need to utilize some psychological strategies (see the next section), and to start to employ them effectively in practice ahead of his return to test cricket (conscious competence). Finally, the goal was for Alan to become highly proficient at managing distractions without conscious attention (unconscious competence).

Developing the focus plans

In two subsequent sessions, Alan and I discussed the nature of his 'scars' and how many of them were factors outside of his immediate control. He agreed that the most important factors that would affect his future performances at test level were under his control and focused on his preparation for the task ahead. Alan decided that these were to prepare himself technically, physically and mentally, in order to score the most runs he could if the opportunity arose. In short, he had to get his focus on the *task* at hand. Subsequently, we created a re-entry focus plan. In our discussions, Alan often talked about the qualities of world class test batsmen and his desire to become one. The player chose to call this work his 'World Class Re-Entry Plan'. Essentially, the plan required the player to think *in the now* and *on his own* performance (Butler, 1996) rather than about the identified distractions. Alan's plan for his first preparation camp with the test squad is presented below.

Alan's world-class re-entry plan for the test squad's first preparation camp

- Focus on me and stay in the present.

- Chance to use this as an opportunity to get back into the pressure situation.

- Play my own game and this is a good chance to test myself.

- Work on start of innings processes.

- Work on defence.

- Let natural strengths happen.

- Play how I want to play.

- Focus on controlling controllables. Don't change myself or my game. I bring plenty of good things to the side.

- If I start to think too much about others and external things I lose focus on myself and start to doubt. Remember my world-class thinking and my processes. These will bring success.

- Media questions: talk about my game and keep it simple. 'I met the challenge of scoring runs to get recalled and I feel good about this opportunity'.

- Use the camp as an opportunity to do some information gathering about practice facilities we will use and the tours in general.

In terms of dealing with the forthcoming touring environment he had reservations about, we also discussed ways of managing and coping with the unfamiliar countries, playing venues, practice facilities and hectic tour schedule that he was going to encounter. A main feature of this planning was to consider what Alan called 'down time'. Alan felt it was imperative to structure his time away from cricket as much as possible, in order to stay mentally and physically fresh enough to consistently perform during an arduous playing and travelling schedule. The list below documents the strategies Alan generated.

Alan's strategies to make the best of a touring environment he may not like

- Look for opportunities to switch off.
- Keep good contact with home: wife and brother.
- Take fly-rod for fishing trips.
- Take computer (take e-mail addresses).
- Take plenty of fishing reading/magazines.
- Don't compromise my values.
- Be self-sufficient.

Alan asked me to type up the lists we had generated, so he had a written record of our discussions to refer back to when he felt he needed to reflect on his preparation and performances.

Grounding

The second intervention with Alan involved a grounding technique (Joyce & Sills, 2001) to address Alan's attention 'wandering' (his own word) when batting. Alan was aware that this 'wandering' tended to happen in breaks in play between deliveries and was characterized by his mind drifting onto past or future events, or uncontrollable factors. Grounding as a technique originates from Gestalt psychology (see for example, Mackewn, 1997) and is a useful technique to bring thoughts and attention to the *present* moment, and I have found it particularly effective to help athletes get in the 'here and now' either when performing or preparing to perform.

To demonstrate this technique to Alan I asked him to be aware of his thoughts, feelings and sensations when I directed him to:

- feel his back against the chair he was sitting on;
- feel his feet on the ground where he was standing;
- hear the noise of the clock ticking in the office;
- notice the different colours he could see in his immediate field of vision.

Alan and I discussed how this grounding technique might be applied to re-focus his thinking and behaviour at the crease between deliveries, and how the technique could quickly be practised and applied on the cricket field, in his net practice, or in any other situation in which he sensed he was being distracted from his present task. Alan grasped the technique and its potential immediately and suggested that he could momentarily 'ground' his feet on the floor (take notice of the feeling of the ground under his feet) at the crease, pay attention to the feel of his bat in his hands, or look at advertising boards around the ground, if he sensed he was getting away from the present moment and being distracted by other internal or external factors.

Over the winter Alan successfully regained his place in the test match side, and through e-mail feedback (see later reflections) it appeared the interventions had worked well. We met again and worked together the following summer when he was contracted again to the English county club. By this time he was also under a 'central' contract with his national cricket board and was asked by his country's test team sport psychologist to complete an off-season goal-setting sheet which Alan asked for my assistance with. Whilst I will not reproduce the nature of the goal-setting within this chapter, assisting Alan with this task gave us the opportunity to strengthen our working relationship.

Toward the latter part of the summer when he was preparing to leave England for his country's home series against the West Indies, Alan asked if we could find time to construct another focus plan for him to build on the successful plan the previous winter. The plan below is the product of two sessions where we discussed key issues in his preparation for a test series against a very different type of opposition, in a rapidly changing team environment with a new coach and captain. The focus of these sessions followed the same process as had been successful previously (i.e. identifying potential distractions and raising self-awareness of solutions; Moran, 1996; Ravizza, 1999). The self-talk he anticipated using appears in capitals.

Alan's focus plan for home series vs West Indies

- Being myself is vital to me feeling at ease with everything. Achieve this through contributing fully in team meetings, not acting like a 'fringe' player, making sure everything is right for my preparation as well as the team's.

- Clear thinking when the pressure is on, achieve this by staying in present (not getting caught up in emotions of the moment, playing for country). Not buying into media stories of my own personal chances or test record or the opposition's strengths and weaknesses. Not getting distracted by others' expectations.

- Phrases like 'play in the moment' and 'businesslike' are good as they are reminders that it is still the same game at test level . . . play the game not the test-match occasion.

- The importance of getting back home and getting some good quality preparation in for domestic cricket (not focusing only on the test series) . . . this will be good practice at staying in the present with plenty of possible media distractions.

- I have a good record against the West Indies . . . I know how to succeed . . . focus on my game plan . . . FIGHT HARD, GET IN RHYTHM are two good focus phrases for me at the crease.

- Importance of not getting ahead of myself. Use grounding to stay in the moment. Take my time batting.

Alan had a successful home series and our work continued by e-mail when Alan prepared in his own country for a winter tour to India. In test cricket India is recognized to be one of the hardest touring environments, with huge stadiums and crowds showing intense, passionate vocal support. In recent years India has been deemed to be one of the most difficult countries in which to win a test series for opposition countries. Below is the culmination of Alan's plans for the potential distractions and the self-talk (in capitals) he anticipated using when in the situation.

Alan's India tour focus plan

Likely 'external' distractions:

- Sea of people.
- Deafening whistles and horns.
- Loud build-ups to key balls (first ball of innings, after dismissals).
- Intense concrete jungle stadiums.

- Sweat dribbling into face and wet hands and shirts.
- Different language and wild mannerisms.

Start of innings processes:

- Use chewy and zinc, have quiet time away from the team to gather my thoughts.

- Good body language walking to middle, marking crease and squatting to adjust eyes to light and understand the noises.

- FIGHT for as long as it takes to feel comfortable. Finish the over even if it means not to score.

- Awareness to be only of time rather than runs. Play my own game and JUST BAT.

- Gloves and drinks will only remind myself to FIGHT.

- Rag in pocket will remind me of grounding at crease.

Batting plans:

- Playing spin: keep looking to score, steal runs, watch the ball hard, play the ball off the wicket, if it's really 'ragging': FIGHT.

- Playing swing: play the ball late, stand and wait, gradually develop shots.

Alan performed excellently on this tour, establishing several batting records, and as a result of his success he was unable to finish his contract with the English county team. However, our work continued by e-mail and he asked for the same planning process to take place for his country's tour to face England in a test match series. Whilst we continued to remain in contact by e-mail for some time after, this proved to be the last intervention with Alan.

3.4 Evaluation of interventions

Partington and Orlick (1987) suggested that it is important to evaluate consultancy for ethical, scientific, educational and professional reasons. Anderson, Miles, Mahoney and Robinson (2002) proposed a framework within which appropriate evaluation for sport psychology interventions might take place. With regard to what could be evaluated, Anderson *et al.*, suggested the following four effectiveness indicators: psychological skills, athlete responses to support, performance, and quality of support.

Psychological skills

Assessment of psychological skills is often undertaken using standardized psychometrics, and these tools can be valuable for monitoring changes (Anderson *et al.*, 2002). However, I have not found questionnaires particularly suitable or specific enough for elite performers, and furthermore other experienced consultants have questioned their usefulness in applied settings (e.g. Bull, 1995; Rotella, 1990). The assessment of Alan and the ongoing monitoring was interview/e-mail based, thus no numerical data was available to plot improvements in psychological skills. Cox (1997) has argued that, to assess and monitor psychological skills, sufficient information can be gained through talking with the athlete without the need for psychometric assessment. Whilst this could be argued to be too subjective, in my experience the complex nature of consultancy, particularly with elite athletes, often means that evaluation of intervention outcomes tends to focus more on athlete responses to the support.

Athlete responses to the support

Vealey (1994) suggested that athletes' responses to sport psychology support can influence the effectiveness of the services and it is thus appropriate to evaluate these responses. These include changes in the client's knowledge of, and attitude towards, sport psychology practice, as well as actual use of mental skills (Anderson *et al.*, 2002).

Whilst this may be a subjective method of evaluation, Alan's informal feedback did give rich information about the intervention meeting his perceived needs. Examples of qualitative feedback from Alan were 'I'm thinking clearer and planning better' and 'I'm more focused at the crease'. Additionally, the strength of the relationship in consultancy work is fundamental to the success of any intervention regardless of the psychological approach taken, and effective working alliances between psychologist and athlete provide the vehicle for behavioural change to take place (Poczwardowski, Sherman & Henschen, 1998). This aspect of sport psychology delivery is often overlooked, with more emphasis given to techniques and skills in the literature. The close relationship that was formed enabled Alan and I to freely discuss and assess the strategies that were being suggested and used, and the ongoing feedback from Alan was very good in this regard. Assessing the quality of a working relationship is not easy; however my indicators of this with Alan were a willingness on his part to disclose highly personal information, he used words such as 'trust you' and there was an openness to have frank discussions about his performances.

Performance

Throughout this chapter I have purposefully omitted Alan's performance statistics during the intervention period (runs scored, averages) in order to preserve his anonymity. Whilst feedback from the previous section's self-report measures demonstrated Alan's high satisfaction and positive view, it is clear it would be erroneous to attribute a causal relationship between the interventions used (focus plans and grounding) and the enhanced cricket performances that Alan produced at test level during the intervention period.

Evidently, many other factors may have contributed to Alan's success, such as the quality of pitches, technical coaching, tactical interventions by the coach/captain and practice time. However, it has been argued that self-report measures are better markers than performance outcomes in assessing effectiveness of interventions in consultancy situations (Murphy, 1995; Ravizza 1988; Rotella, 1990), hence, in this respect, the focus plans and grounding appeared successful in helping Alan manage some of the distractions he felt could affect him and his performance on and off the field.

Quality of the support

A formal approach to evaluating consultant effectiveness is through the use of standardized feedback forms (Anderson *et al.*, 2002). To this effect, a cricket-specific Consultant Evaluation Form (Partington & Orlick, 1987) was also e-mailed to Alan after his tour to England. Feedback from this form was extremely positive, with top ratings on categories relating to the types of intervention used and their impact, and on consultant characteristics (e.g. trustworthiness, flexibility, easy to relate to).

Anderson *et al.*, (2002) have also suggested that the quality of the support could also include an evaluation of consultant effectiveness through reflective practice. The following section briefly presents a personal reflection on the quality of the support provided for Alan.

3.5 Evaluation of consultant effectiveness/reflective practice

The work that has been described is based upon a top-down, evidence-based approach (Weinberg & Gould, 2003); therefore the consultancy has been based upon social scientific literature. In order to bridge the gap between knowing, articulating and using this knowledge, I have adopted Johns's (2000) model of reflective practice. This model suggests that reflective practice should involve these types of questions:

- What happened?
- What were you thinking and feeling?
- What was good and bad about the experience?
- What sense did you make of the situation?

- What else could have been done?
- If it arose again what would you do?

Cultural aspects

Knowles, Gilbourne and Tomlinson (2007) suggest that personal development occurs through reflection. I personally feel I developed through my work with Alan in a number of ways. It was a great opportunity to work with a high-profile overseas player and therefore adapt to the cultural differences that were evident. I have conducted much consultancy work within motor racing, where I have had experience of dealing with different nationalities, and I place great emphasis on gaining an understanding of the different sport-specific terminology that is used by athletes. I feel this is an important task for rapport building by consultants who find themselves working with athletes from different nations.

High-profile player

The opportunity to work with a high profile player was challenging. The opportunity to consult with a leading player in the world game did seem to add importance to the work and with that came a sense of pressure. I raised this issue within my monthly peer supervision with an experienced colleague working in elite sport and found that the support and advice offered greatly diminished the feeling that I had. I believe that consultants can erroneously feel that high-profile athletes will require 'special' interventions, when in reality they, like others, may simply need space and time in a trusted relationship to explore their thinking and behaviours. For the past 10 years I have found time to reflect on my practice in this 'safe' environment with a trusted colleague, and believe that my consultancy skills have benefited enormously as a result. I would encourage any sport psychologist to engage in ongoing supervision post-qualification for continued professional development.

Strong relationships

I also believe the strong relationship developed with Alan led to a collaborative approach to how the interventions needed to be developed and delivered. This work strengthened my belief in the importance of credibility and trust in consultancy work. The counselling literature (e.g. Salacuse) has emphasized the need to build a good working relationship through two-way communication, genuine commitment, reliability and respect for the client as a person. Moreover, Poczwardowski *et al.,* (1998) have argued that the quality of psychological support delivery cannot be seen as separate from the professional consulting relationship, with the consultant–athlete relationship being a means by which a consultant can develop changes in athlete behaviour. I feel this emphasizes the importance of the professional relationship, and on reflection I feel this is one of my strongest points as a consultant. In this regard, the importance of sound counselling and listening skills in sport psychology consultancy should not be underestimated (Hemmings, 1999).

Furthermore, the work emphasized that sport psychology service delivery is not all about teaching techniques and skills. The work over the two-year period demonstrated the need for me to be available only when 'needed' by Alan. There were plenty of other times when I did not feel I was engaging in those types of tasks with Alan, nor was it necessary for me to do so. Andersen (2000) has written much valuable literature about the processes of delivery, and the consultant can do much by 'just being there' with athletes. Reflecting on the experience with Alan, it further emphasized there is as much need in consultancy work to take opportunities to build rapport with athletes as there is to be concerned with delivering practical skills. I believe the two aspects are of equal importance.

Means of communication

One particularly challenging aspect of the work was the mechanism for communication between Alan and I when we were in different countries from one another. Whilst e-mail is positive in that it allows for frequent contact, I found it quite limited when it was the sole means of communication over a winter period. One particular problem with the Internet is the absence of face-to-face contact and the non-verbal cues such as body language (see Watson, Tenebaum, Lido & Alfermann, 2001; Zizzi & Perna, 2002) which are normally available. In the work with Alan when he was overseas, all contact was via e-mail and I was concerned about its effectiveness. I relayed these concerns to Alan, and we decided that he and I should use capitals and bold typeface within e-mails to emphasize what we felt were key issues.

Time constraints

The time available for individual intervention during consultancy work with the cricket club may have impacted on the quality of the support. The need for good listening and reflection in the first sessions was vital in establishing rapport, and the work with Alan emphasized the need for flexibility and sensitivity when re-assessing his needs after his first overseas tours. Whilst the time-limited sessions sometimes created situations where I would have preferred to spend longer with Alan, I also feel that, to some degree, the small time available enabled very brief, solution-focused strategies to be created and acted upon. Again, my monthly peer supervision proved very beneficial in this regard. Discussing this time pressure in support work and ways to manage it was useful in keeping me focused on the best use of the time I had available. For instance, simply clarifying with Alan 'what he wanted to get out of the time available' at the start of each session ensured we kept to that issue when time was short.

Practice-based evidence

Whilst adhering to evidence-based practice within my current professional boundaries, there were occasions when relevant literature was not available to inform my practice (e.g. specific examples of the grounding technique being used in cricket). In this situation, as suggested by Anderson, Knowles and Gilbourne (2004), practice-based evidence acquired through reflective practice assumed importance in my work with Alan.

Sharing concerns

Finally, I found that Alan employed the deepest planning of performance issues that I had dealt with in my professional work. The lengths to which Alan would probe potential difficulties and distractions sometimes made me concerned that over-analysis may have been taking place. I was cautious here and shared this with him. On reflection, this sharing was useful and I also believe that the heightened levels of self-awareness and planning Alan showed were fundamental in the success of the work. I also feel that this awareness contributed to the very simple, practical strategies that were identified, and the simplicity of the work was a vital factor in its success.

3.6 Summary

The ability to control thought processes, to concentrate on a task, is widely recognized as important for effective sports performance (Nideffer, 1999). Interventions used by applied sport psychologists often involve assisting athletes to identify task-relevant aspects of their performance and also to develop the ability to recognize and deal with distractions.

This chapter has described intervention work with an international test cricketer over a two-year period. The cricketer had experienced a poor test career prior to intervention (being dropped twice) and talked of the 'scars' of previous performances. The needs analysis (interviewing cricketer about previous experiences and current thoughts/behaviours) clarified the types of distractions that the cricketer felt he faced.

The intervention phase demonstrated the usefulness of raising awareness of potential distractions, and then described the implementation of a combination of focus plans and a grounding technique to manage those distractions. Reflections on the effectiveness of the work emphasized the importance of the psychologist–athlete relationship as a vehicle for behavioural change.

Questions for students

1 Identify some internal and external distractions in cricket/other sports.

2 Discuss the unique settings in which sport psychologists operate compared with more traditional counselling/clinical psychology settings.

3 Apply the four-step awareness model used in this chapter to another sport performance issue.

4 Reflect on how the grounding technique described in this chapter could be applied to other sports.

5 Discuss how the athlete–psychologist relationship can be strengthened to bring about positive change in athlete behaviour.

References

Andersen, M. (2000) *Doing Sport Psychology*. Human Kinetics, Champaign, IL.

Andersen, M. (2005) *Sport Psychology in Practice*. Human Kinetics, Champaign, IL.

Anderson, A., Knowles, Z. and Gilbourne, D. (2004) Reflective practice for sport psychologists: Concepts, models, practical implications, and thoughts on dissemination. *The Sport Psychologist* **18**, 188–203.

Anderson, A., Miles, A., Mahoney, C. *et al.* (2002) Evaluating the effectiveness of applied sport psychology practice: Making the case for a case study approach. *The Sport Psychologist* **16**, 432–453.

Boutcher, S. and Rotella, R. (1987) A psychological skills education program for closed skill performance enhancement. *The Sport Psychologist* **1**, 127–137.

Bull, S. (1995) Reflections on a 5-year consultancy program with the England women's cricket team. *The Sport Psychologist* **9**, 148–163.

Bull, S. (1997) The immersion approach. In: Butler, R., (ed.), *Sport Psychology in Performance*, pp. 177–202. Butterworth-Heinemann, Oxford.

Bull, S., Shambrook, C., James, W. *et al.* (2005) Towards an understanding of mental toughness in elite English cricketers. *Journal of Applied Sport Psychology* **17**, 209–227.

Butler, R. (1989) Psychological preparation for Olympic boxers. In: Kremer, J. and Crawford, W. (eds), *The Psychology of Sport: Theory and Practice*, pp. 74–84. British Psychological Society, Leicester.

Butler, R. (1996) *Sport Psychology in Action*. Routledge, London.

Carron, B., Hausenblaus, H. and Eys, M. (2002) *Group Dynamics in Sport.* Fitness Information Technology, Morgantown, WV.

Cox, R. (1997) The individual consultation: the fall and rise of a professional golfer. In: Butler, R., (ed.), *Sport Psychology in Performance*, pp. 129–146. Butterworth-Heinemann, Oxford.

Dryden, W. (2006) *Counselling in a Nutshell.* Sage, London.

Gordon, S. (1992) Concentration skills for bowling in cricket. *Sports Coach* **15**, 34–39.

Halliwell, W. (1990). Providing sport psychology consulting services to a professional sport organization. *The Sport Psychologist* **4**, 369–377.

Hardy, L., Jones, G. and Gould, D. (1996) *Understanding Psychological Preparation for Sport: Theory and Practice of Elite Performers.* Wiley, Chichester.

Hardy, L. and Parfitt, G. (1994) The development of a model for the provision of psychological support to a national squad. *The Sport Psychologist* **8**, 126–142.

Hemmings, B. (1999) Making contact in consultancy work: laying the foundations for successful intervention. In: Steinberg, H. and Cockerill, I., (eds), *Sport Psychology in Practice: The Early Stages*, pp. 16–23. British Psychological Society, Leicester.

Johns, C. (2000) *Becoming a Reflective Practitioner: A Reflective and Holistic Approach to Clinical Nursing, Practice Development and Clinical Supervision.* Blackwell, Oxford.

Joyce, P. and Sills, C. (2001) *Skills in Gestalt Counselling and Psychotherapy.* Sage, London.

Knowles, Z., Gilbourne, D. and Tomlinson, V. (2007) Reflections on the application of reflective practice for supervision in applied sport psychology. *The Sport Psychologist* **21**, 109–122.

Mackewn, J. (1997). *Developing Gestalt Counselling.* Sage, London.

May, J. and Veach, T. (1987) The U.S. Alpine Ski team psychology programme: a proposed consultancy model. In: May, J. and Asken, M. (eds), *Sport Psychology: The Psychological Health of the Athlete*, pp. 19–39. PMA, New York.

May, T. (2001) *Social Research: Issues, Methods and Process.* Open University Press, Buckingham.

McMahon, G. (2001) *Confidence Works: Learn to be Your Own Life Coach.* Sheldon, London.

McMullin, R. (2000) *The New Handbook of Cognitive Therapy Techniques.* Norton, London.

Moran, A. (1996) *The Psychology of Concentration in Sport Performers: A Cognitive Analysis.* Psychology Press, London.

Müller, S. and Abernethy, B. (2005) Skill learning from an expertise perspective: issues and implications for practice and coaching in cricket. In: Dosil, J. (ed.), *The Sport Psychologist's Handbook: A Guide for Sport-Specific Performance Enhancement*, pp. 245–264. Wiley, Chichester.

Müller, S., Abernethy, B. and Farrow, D. (2006) How do world-class batsmen anticipate a bowler's intentions? *Quarterly Journal of Experimental Psychology* **59**, 2162–2186.

Munroe, K., Giacobbi, P., Hall, C. *et al.* (2000) The four W's of imagery use: where, when, why and what. *The Sport Psychologist* **14**, 119–137.

Murphy, S. (1995) *Sport Psychology Interventions.* Human Kinetics, Champaign, IL.

Nelson, J., Duncan, C. and Kiecker, P. (1993) Toward an understanding of the distraction construct in marketing. *Journal of Business Research* **26**, 201–221.

Nideffer, R. (1999) Attention and concentration control strategies. In: Williams, J., (ed.), *Applied Sport Psychology: Personal Growth to Peak Performance*, pp. 243–261. Mayfield Press, Toronto.

Orlick, T. (1990) *In Pursuit of Excellence.* Leisure Press, Illinois.

Palmer, S. and Dryden, W. (1995) *Counselling for Stress Problems.* Sage, London.

Partington, J. and Orlick, T. (1987) The sport psychology consultant evaluation form. *The Sport Psychologist* **1**, 309–317.

Poczwardowski, A., Sherman, C. and Henschen, K. (1998) A sport psychology services delivery heuristic: building on theory and practice. *The Sport Psychologist* **12**, 191–207.

Ravizza, K. (1988) Gaining entry with athletic personnel for season-long consulting. *The Sport Psychologist* **2**, 243–254.

Ravizza, K. (1999) Increasing awareness for sports performance. In: Williams, J. (ed.), *Applied Sport Psychology: Personal Growth to Peak Performance*, pp. 171–181. Mayfield Press, Toronto.

Rotella, R. (1990) Providing sport psychology consulting services to professional athletes. *The Sport Psychologist* **4**, 409–417.

Salacuse, J. (1994) *The Art of Advice.* Times Books, New York.

Taylor, J. (1995) A conceptual model for integrating athletes' needs and sport demands in the development of competitive mental preparation strategies. *The Sport Psychologist* **9**, 339–357.

Taylor, J. and Wilson, S. (2005) *Applying Sport Psychology.* Human Kinetics, Champaign, IL.

Theodorakis, Y., Weinberg, R., Natis, P., *et al.* (2000). The effects of motivational versus instructional self-talk on improving motor performance. *The Sport Psychologist* **14**, 253–272.

Thomas, P. (1990). *An Overview of the Performance Enhancement Process in Applied Psychology.* US Olympic Center, Colorado Springs.

Vealey, R. (1994) Current status and prominent issues in sport psychology intervention. *Medicine and Science in Sport and Exercise* **26**, 495–502.

Watson, J., Tenebaum, G., Lido, R. *et al.* (2001) ISSP position stand on the use of the internet in sport psychology. *ISSP Newsletter* **9**, 14–21.

Weinberg, R., Burton, D., Yukelson, D. *et al.* (2000). Perceived goal-setting practices of Olympic athletes: an exploratory investigation. *The Sport Psychologist* **14**, 279–295.

Weinberg, R. and Gould, D. (2003). *Foundations of Sport and Exercise Psychology* (3rd edn). Human Kinetics, Champaign, IL.

Weinberg, R. and Williams, J. (1999) Integrating and implementing a psychological skills training program. In: Williams, J. (ed.), *Applied Sport Psychology: Personal Growth to Peak Performance*, pp. 274–298. Mayfield Press, Toronto.

Zizzi, S. and Perna, F. (2002) Integrating web pages and e-mail into sport psychology consultations. *The Sport Psychologist* **16**, 416–431.

4

Consultancy in the Ring: Psychological Support to a World Champion Professional Boxer

Andrew M. Lane

University of Wolverhampton, Wolverhampton, UK

4.1 Introduction/background information

The career of a professional boxer is a challenging and potentially stressful one. There can only be one winner of a boxing contest. Whilst the 'winner takes all' idea is a characteristic common to many sports, boxing is different in a number of ways (see Oates, 1987). The nature of boxing means that competitors can sometimes aim to injure each other during competition; winning can occur by injuring your opponent more that he/she can injure you. The career structure in professional boxing exacerbates the importance of winning. Winners can go on to higher paydays, while losers become journeymen or opponents for up-and-coming boxers. A career can change direction on the basis of a single contest, and boxers are paid only when they fight.[1] Stories of professional fighters earning and spending millions of pounds are common (Oates,

[1] There are examples where boxers sign deals for a number of fights. Boxers typically work on fight-by-fight basis.

Applied Sport Psychology   Edited by Brian Hemmings and Tim Holder
© 2009 John Wiley & Sons, Ltd

1987). While athletes in many sports play their game, boxers 'fight' (see Hatton, 2007 for an autobiographical example) and boxing is a challenging performance culture to operate in.

The work described in this chapter involves a professional boxer's preparation for a World Championship contest. When our work started, he had fought 33 professional contests, winning 30 and losing three. His defeats included a high-profile world title fight and also losses against opponents that, in his own words, he should have beaten. He had recently returned to boxing after a two-year absence and his next fight was against a boxer with a record of 40 wins (38 knockouts) from 45 fights.

An initial meeting had been arranged by the boxer's coach, who was a mature student at the university where I was lecturing and knew that I had a boxing experience (see later reflections). During our initial discussions, I described how I planned to work, emphasizing that the boxer was likely to already have a number of desirable psychological characteristics. I explained that it would be unlikely that he could have attained his current level of performance without possessing such qualities, and what I aimed to achieve was to provide strategies to enhance these skills. Interestingly, canvassing the opinions of elite athletes has been used to develop the qualities required to be an elite performer, particularly the concept of mental toughness (Jones, Hanton and Connaughton, 2002).

4.2 Initial needs assessment

Performance profile

A performance profile was developed to ascertain the self-rated strengths and weaknesses of the boxer (see Butler and Hardy, 1992, Butler, Smith and Irwin, 1993). A second profile was developed later that assessed the coach's perceptions of the boxer's strengths and weaknesses. The theoretical basis for performance profiling derives from personal construct theory (Kelly, 1955). Personal construct theory proposes that each individual has a unique style through which he/she makes sense of his or her environment. Insight into this relationship is conducted by knowing the central tenets of personal style (see Weston, 2008 for an explanation). A key point to the effectiveness of personal construct theory as the basis of intervention techniques to change behaviour is that the individual becomes aware of their personal constructs.

The performance profile was constructed by asking the boxer to think about what constituted an ideal boxing performance and which boxers exemplified those qualities. To facilitate the process, I asked the boxer to describe the qualities of his role models. I have found it is easier for athletes to describe someone they know and list his or her strengths and weaknesses, rather than to try to describe an 'ideal' performance.

The conversation on who were elite boxers started and flowed easily, identifying several world champions and the qualities they demonstrated in competition. What emerged from this conversation was that he had worked as a sparring partner for several different world champions. One sparring partner was a multi-weight world champion and another had made over 15 successful defences. At the time, the boxer had experience of working with many successful boxers and therefore, alongside his previous world championship challenge, he had a reasonable understanding of what was required to achieve his goal of winning the world championship.

The process of identifying the personal constructs was an iterative process. He was asked to identify and describe why one boxer was superior to a different boxer, why two successful boxers were alike and why two unsuccessful boxers were alike. The process involved probing the boxer to elaborate on differences and similarities. A strategy to explore whether he could use these constructs to differentiate a winning from a losing performance was conducted by asking him to apply these constructs for future contests, for example, 'Who would win if boxer A competed against boxer B, and why?'

This approach to identifying the qualities of the elite performer is slightly different from how performance profiling has previously been conducted when athletes identify constructs in a group and 'brainstorm' ideas (e.g. Butler *et al.*, 1993). A fundamental part of personal construct theory is that each individual has unique constructs and these guide how information is interpreted. Therefore, practitioners should seek to identify the uniqueness of these qualities using an ideographic (within-subject) approach.

The boxer identified a range of personal constructs or qualities. Of the 13 qualities identified, eight related to technical skills, one quality was psychological and four qualities were physiological (see Figure 4.1). It is worth emphasizing that the constructs are personal and might not generalize to other boxers. The constructs relate to his style of boxing, and styles between boxers can vary hugely. What was clear from this discussion was that he used these constructs to discriminate good performance from poor performance. Following this, he rated a number of elite boxers using a 10-point scale anchored by 'not at all' (1) and 'very much so' (10), before rating himself on the same scale. As seen in Figure 4.1, he rated himself considerably lower than the 'ideal boxer' he selected for comparison.

I telephoned the coach to discuss the profiling process and to ask the coach to rate the boxer on the same constructs. Comparing the athlete's self-perception and the coach's perception of the athlete is particularly relevant in boxing. Boxers work extensively with their coaches in training and in pre-bout warm-up, and receive feedback from their corner at the end of each round. Hence, if the coach is aware of the personal constructs that the boxer uses to define elite performance, then he can shape their training and feedback accordingly. Research has identified that enhanced communication is a benefit of the performance profiling process, particularly between athletes and coaches (e.g. Dale and Wrisberg, 1996; Weston, 2008). Using the same qualities identified

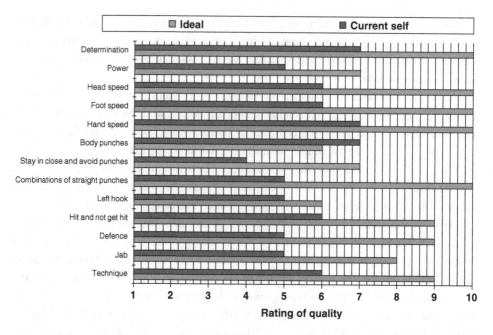

Figure 4.1 Performance profile: current self vs ideal boxer.

by the boxer, the coach provided his ratings. The initial discussion with the coach involved explaining to the coach the constructs the boxer had identified. The coach queried these, which in itself is a useful process as the coach was being made aware of the athlete's perception. As Figure 4.2 illustrates, the coach provided higher ratings on the constructs than those given by the boxer. Following this conversation, the coach agreed that we should set up a three-way conversation (coach/boxer/psychologist) on the results of the profiles to plan a way forward.

To summarize the assessment, the boxer appeared to have good knowledge of what constituted the 'qualities' of a world champion. However his rating of himself on those qualities needed to improve, and these ratings were also lower than those provided by his personal coach.

4.3 Intervention and monitoring

I used a combination of strategies to try to bridge the gap between the 'ideal' self and 'current' self. This included various psychological skills and pre-competition strategies; however in the present chapter I will focus on the use of a videotape analysis intervention.

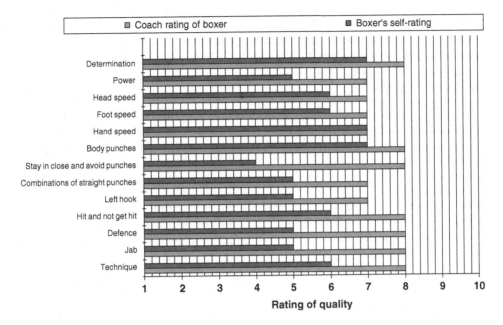

Figure 4.2 Performance profile: coach and boxer comparison.

Using videotape to bridge the difference between current self and ideal self

Recent research has emphasized the value of using modern technology to aid intervention work (Bertram, Cote, Brown, Guadagnoli and Palomaria, 2007; Ives, Straub and Shelley, 2002). Recording an athlete's performance is relatively straightforward with modern technology. When recording athlete performance, careful consideration is needed on how it can be used, otherwise the consultant can be overwhelmed with data. Sessions were videotaped around key aspects of performance, including sparring and circuit training. In terms of viewing the sparring footage, the coach, the boxer and I viewed the taped session on a large projector screen. We viewed the videotape, stopping it at key moments to discuss relevant issues. These conversations were productive, with coach and boxer being receptive to each other's opinion. There were differences in opinion, sometimes heated, but they served to intensify both the boxer's and the coach's reflections on what went well and what needed improving. The boxer's explanation of his performance reaffirmed the concepts identified in the performance profile. These discussions led to setting specific goals for future sessions. My role was to facilitate discussion and also show successful performance in training and link this with the constructs identified in the performance profile.

The training regime had two key sparring sessions per week. The boxer would be relatively rested for each of these key sparring sessions. Sparring partners were hired to simulate the opponent's style. A great deal of planning was involved for each of these sessions. We watched taped performances of the next sparring partner to develop plans for the spar. By using simple performance analysis techniques such as counting the number of punches typically thrown in each round and when they are thrown, we identified common patterns for each boxer. For example, a boxer might throw more punches at the start of a round, or move to one side more than the other. We developed a series of 'if–then' plans (Achtziger, Gollwitzer, and Sheeran, 2008); for example if the opponent came out fast, then the plan was to spoil or move away, but not get into a 'battle'. The boxer would wait for the fast start to slow down, and then look to take control in the second half of the round if this plan was deemed appropriate. Recent research has supported the utility of if–then plans to assist an individual in making desirable self-regulatory decisions. For example, Achtziger *et al.*, found that tennis players who used if–then plans reported lower levels of state anxiety and improved performance.

Viewing videotaped sessions also aided usage of psychological skills such as imagery (see Smith and Wright, 2008). If after watching a sequence from a sparring session, the boxer saw an opportunity when he should have stayed close and avoided punches, looking to land short punches such as a left hook, he would repeat the sequence where this occurred, and then image himself performing the desired move. This process would be repeated not only when he identified parts of performance that he felt warranted changing, but also when he identified successful performance. Imaging successful performance was an important way of reinforcing feelings of success (Bandura, 1997). The act of identifying success by asking the boxer to develop clear images of repeating successful performance was found to be a useful strategy shortly before the contest when thoughts and feelings that accompany self-doubt came in. For example, on the evening of the world championship when reiterating plans for the contest, he was less confident of his capability to enact these than when we went through them previously. I was able to remind the boxer of times when he had performed well and used imagery as a reinforcement strategy. Video analysis was seen as an important within training, partly because it was novel for the boxer to watch, so he remembered these sessions clearly.

In addition to enhancing the perceptions of the qualities identified in the performance profile, watching videotaped sparring sessions and the ensuing discussions also enhanced coach–athlete communication. The coach indicated that he felt that the boxer did not always use advice given in the corner between rounds. There are a number of reasons why this could have been. The advice could have been deliberately ignored; it could have been misheard; or it could be that the boxer tried to act on the advice but was not able to do so because he was prevented by his opponent. The coach thought that this was an issue worth exploring in order to find a solution.

Videotaped sparring sessions provided both coach and boxer with a different perspective on the coach feedback process. The boxer explained what he was trying to do during the spar, and the coach reflected and commented on this information. What emerged was that the coach and boxer had not previously engaged in such frank discussions and this aided coach–boxer communication. Videotape removed biases from memory; both coach and athlete looked at the actual performance rather than a memory of the performance. This was important in terms of the effectiveness of the exercise. I question whether we would have made such progress without videotape evidence. Coach and boxer offered insight into what they heard and thought they were doing at the time. There were occasions when considerable disparities emerged and discussion of these differences seemed to help boxer and coach understand each other better.

For example, the coach once stopped the videotape to discuss a situation where he had just given advice between rounds, and had told the boxer to keep his right hand up as his opponent was looking to land a left hook. Once the round started, within 10 seconds, the boxer's hands were held low. Videotape enabled us to listen to the advice given between rounds several times for reasons of accuracy. The coach and I could ask the boxer what information he had heard and how he had planned to use this information. He understood the instruction to 'keep your hands up', as meaning make sure you pay attention to defence and look to avoid his left hook. He indicated that he was aware that his opponent was looking to land the left hook, and therefore had developed plans to either block it or move away. When watching the videotape, he agreed with the coach that the opponent was looking to land the left hook; however he perceived the tactic/skill to counteract this differently from the coach. It should be noted that the boxer's explanation was consistent with the constructs he identified in the performance profile. He identified defence as a key construct, whereas the coach conceptualized this construct differently. When the coach spoke about defence, he tended to focus on specific skills such as holding the hands high and specific blocks, whereas the boxer perceived defence to be more about reflexes and the ability to identify when the punches would be thrown and take evasive action. The above example illustrates the usefulness of using videotape sparring sessions to enhance the coach–athlete relationship.

Circuit training sessions were videotaped to focus on movement quality when performing maximally. Performance profile qualities being targeted included speed, power and determination. We had discussed these qualities in the development of the performance profile and we monitored the number of repetitions performed on each circuit training 'station' and used this score as the basis for future goals. We also monitored how relaxed he looked, as he wished to avoid looking tired. Boxers typically do not want to present a message to their opponent that they are feeling tired. Lane (2007) reported that perceptions of fatigue are influenced by other mood states experienced simultaneously. If athletes feel fatigued and depressed, then they tend to also feel angry, leading to thoughts of despair and potential withdrawal from the activity. If athletes feel fatigue and happy simultaneously, then fatigue is explained as a necessary part of goal attainment. It is argued that upholding positive self-talk is crucial to interpreting fatigue as manageable; hence the boxer and I discussed mood-management strategies to achieve this (see Lane, 2007).

4.4 Evaluation of intervention

Evaluating the extent to which the intervention was successful is complicated by difficulties in being able to extricate the role of sport psychology support in this process. We repeated the performance profile (see Figure 4.3) before the World Championship Contest. As Figure 4.3 indicates, the boxer reported meaningful improvement in the personal constructs identified, and the ensuing discussion in which he provided explanation justifying why he believed he had improved was a useful exercise as it acted as a performance 'reminder'. In terms of the specific skills, he showed improvements in technical skills such as head speed, foot speed, staying in close and avoiding punches, combinations of straight punches, left hook, hitting and not getting hit, defence, jab and technique. He showed the largest difference in his perception of determination.

It would be easy to evaluate the effectiveness of the intervention by suggesting it was effective because he won the contest related to this work. Caution is urged

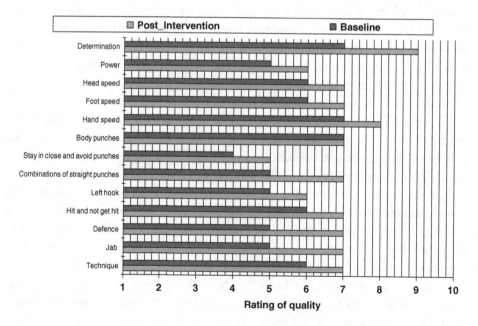

Figure 4.3 Performance profile post-intervention assessment.

before making such an assumption. The boxer's success was due to his ability and decision-making on the night of the fight. In the ring, he made the decisions on which punch to throw, when to attack, when to block and when to defend. I feel my contribution was to increase the likelihood that he would be prepared for the competition and be in a position to make the best decisions he could on the day of the competition. He showed improvements in qualities identified in the performance profile and he reported feeling a greater sense of self-confidence.

4.5 Evaluation of consultant effectiveness/refelective practice

Recent research has indicated difficulties that sport psychologists can face when working with professional athletes. Despite the proposed benefits of interventions, athletes remain cautious (see Thelwell, 2008). The credibility of the applied work was possibly increased because the coach was central to the programme and it is worth reflecting on why the coach and I could work effectively together. He had 30 years of practical experience and was beginning a sport science degree, and therefore was starting to acquire formal qualifications to compliment his experience. I had boxing experiences as an athlete and later as a coach (I set up the boxing club at the university and coached for two years at an amateur club), complemented by several years of academic study. In my work as a sport psychologist, I feel that it is important to emphasize the value of perceiving that the coach is supportive of your work and I have documented issues related to gaining credibility in boxing (see Lane, 2006). The coach's acceptance of an academic working in the boxing team is likely to have been helped by his decision to study for a degree. There are few academics involved in boxing and few boxing coaches have university qualifications.

Credibility

Developing credibility was highly important and I believe that I started from a favourable position. I emphasize that this statement is my perception and a construction of reality from my perspective. This is, however, a perception that I question when I reflect back on this work, but one I still believe holds true. Initial credibility was fortuitous and based on experiences not easily included in training programmes of aspiring sport psychologists. Perceptions of sport psychology can vary and this phenomenon applies across sports. Some boxing gyms have adverts from 'sport psychologists' displayed on their walls, and this is knowledge that I gleaned from personal experiences as a boxer. I had an awareness of what athletes felt about such people and, when going to work with boxers, reflected on such thoughts and how I would work if I met 'myself' several years earlier. At this point, it is worth pointing out that a number of people practise as a sport psychologist who are not suitably qualified (i.e. are not chartered or accredited by the UK professional bodies described in the Introduction to the book). Unqualified people offer no formal quality assurance procedures and this poses a difficulty as coaches and boxers may have mixed experiences of what a sport psychologist 'does'. Caution was needed in exploring attitudes to some of the strategies I could suggest using, something I was particularly aware of in the early stages of the consultancy. For some people, sport psychologists are mystical characters who fix athletes' problems as if they are magicians. My work would never come across in this way. I intentionally strive for improvements in the clients who I work with, but I want them to feel responsible for these changes. My clients will need to work at making changes, they will need to practise, and if they practise appropriately, they should improve. I support the process, but the clients are responsible for improvements. The client should feel responsible for this change, attributing some changes internally to skill and effort. This model of working is unlikely to have me feted as a 'guru'-type figure, as can sometimes occur with non-qualified 'psychologists' in professional sport.

My strategy to develop credibility was to immerse myself in the training environment and to get to know the boxer as much as possible. It is important to know the athlete's psychological strengths and limitations not only as an athlete, but also as a person. Hence, I attended training sessions in boxing gyms across London and adopted diverse roles. I was the driver; the timekeeper; the filler of water bottles; and a training partner on long runs where we would talk. It was during these one-to-one sessions, when interruption was minimal, where a great deal of work went on. My key point here is not to focus on the nature of these specific tasks, but to emphasize that sport science support involves supporting the athlete and this often involves doing mundane tasks. Indeed, Andersen (2006) has emphasized the importance of supporting the athlete in various ways. There are tasks that someone needs to do, and it helps everyone if these are done immediately. Clearly there are some tasks that fall within the remit of a sport psychologist and I should be free to do these tasks. However, there is often time available and I was comfortable with supporting the team even if this involved doing mundane tasks. It seems common sense that people will accept you more if you show a willingness to get involved. Ultimately, you risk losing respect if you give the impression that some tasks are beneath you. During all the individual time we had together (e.g. whilst driving or running), I gathered the boxer's thoughts on training, fighting, attitudes to weight-making, previous experiences and how he coped with the pressures of being a professional boxer.

Dealing with 'myths'

The coach and I would meet or speak on the telephone. During the first month from the initial meeting, I developed good working relationships with the inner circle of the boxer–coach team. My relationships with other boxers, coaches and promoters were not always so straightforward. In certain situations, I deliberately kept quiet in circles where the coach was not the most important figure. It was not always easy to keep quiet. For example, on one occasion, I listened to a conversation on 'making the weight' in which a former world champion suggested that boxers could lose a stone in fat by eating lemons only, with the lemons supposedly dissolving the fat. Clearly, this information is scientifically inaccurate; however the boxer on the lemon-only diet won the contest. Boxing is full of stories of people being able to rapidly lose and regain weight and supposedly maintain strength. It is easy to see why such myths perpetuate. If two boxers compete both having de-hydrated significantly to make weight, one boxer will win (Hall and Lane, 2001). It is likely that the winning boxer will believe that he or she retained strength, whilst the losing boxer might attribute defeat to the 'battle with the scales'. Boxing is rife with such stories. During such times, I would listen and later probe the boxer's view on the issue. I would encourage the coach and boxer to discuss such issues if the boxer was beginning to believe such myths as a usable strategy. Whilst there are many stories of successful weight loss, most boxers know stories of unsuccessful weight loss. Listening to the boxer describe the unsuccessful strategies was an effective way of ensuring that the successful stories were at least questioned.

Having a sport psychologist in the team was a potentially thorny issue throughout. This was especially relevant when we were at boxing tournaments when we would meet other coaches, boxers and the media. When I wished to avoid providing an explanation, I would introduce myself as one of the coaches. The work undertaken within the small unit of the coach and boxer could be easily undone through interactions with other coaches, fighters and promoters. There was a plethora of people offering advice. There were instances during my involvement when the boxer received advice from potentially influential people that contrasted with the work I was doing. At such times, it was important to think carefully on how best to maintain my credibility, and my strategy was to explore the boxer's perception of the content of these types of discussion. By listening to how the boxer interpreted the information, it furthered my understanding of how the boxer thought, and in doing so, started to develop 'in-roads' into how I might be able to tailor my message so that it was meaningful to the boxer. It was therefore important to consider how I 'came across', and to manage the perception by the boxer of what sport psychology involved.

Sport-specific knowledge and experience

At this point, it is worth considering whether my belief that I had credibility was based on my experiences as a boxer. This raises the question whether my belief that I was relatively effective is part of a self-serving bias where I am selecting information that supports this view, and ignoring other information that could be valuable. However I do believe my experience as a boxer was extremely helpful. I believe I organized this information to provide a 'map of the territory', which included having some understanding of the pressures on the fighter, the importance of the coach–athlete interaction and the intensity of the sport. From this position, I felt more confident in my interactions with people in the sport. A question that stems from this line of thinking is whether consultants should have detailed experience of the sport in which they wish to practise. The answer to this question is not straightforward, and I would not wish the reader to think that I am suggesting that sport psychologists should first compete in the sport in which they work. However, I believe sport psychologists should consider consultancy work in sports or with athletes where they think they can be effective. I felt my previous boxing experiences were helpful, and therefore I drew a great deal of confidence from these experiences. It is important for me to introspect and identify information that influences my self-confidence to work effectively as a practitioner.

Sustaining positive relationships in professional boxing can be difficult. Experience and success in boxing is at the top of the hierarchy of knowledge in boxing. Academic qualifications are not typically highly valued and even my experiences as an amateur boxer could be counter-productive. I needed to use my knowledge of the sport to identify where I could be effective and consequently most of the work focused on structuring training sessions so that the boxer maximized the positive effects. Examples of this were modifying how the coach provided feedback between rounds; using performance profiling to identify qualities both the coach and boxer perceived as important; and using videotape to enhance these qualities. In this intervention, changes to what the coach and boxer did were relatively minimal. The boxer already set goals and engaged in some use of imagery. What I felt I achieved was to provide some structure to these techniques.

4.6 Summary

Professional boxing represents a challenging environment in which to work. Performance profiling was used to identify constructs associated with performance. The boxer, coach and sport psychologist viewed videotaped sessions with goal setting stemming from differences between observed and ideal states, and imagery being used to enact this performance symbolically. Coach–boxer communication was developed, which led to a range of benefits such as improved between-round feedback and more specific fight plans. Pre-contest performance profiling data indicated meaningful improvements in constructs identified by the boxer. Credibility amongst the support team was a constant challenge due to the performance environment and influences that exist within professional boxing.

Questions for students

1 What strategies were used to develop credibility with the professional boxer?

2 Discuss the approach to using performance profiling used in this chapter.

3 The present case study used videotape to record the boxer's performance and provide feedback. Consider recent technological developments and propose how these could be used in applied sport psychology.

4 A boxer you are working with is preparing for a world championship fight. What key aspects from his chapter might you use to assist with their psychological preparation?

5 Performance profile data indicated that the coach and athlete identified different constructs as imperative for performance. What strategies might you use to bridge differences in coach–athlete perceptions?

References

Achtziger, A., Gollwitzer, P.M., and Sheeran, P. (2008) Implementation intentions and shielding goal striving from unwanted thoughts and feelings. *Personality and Social Psychology Bulletin* **34**, 381–393.

Andersen, M.B. (2006) It's all about sport performance . . . and something else. In: Dosil, J. (ed.), *The Sport Psychologist's Handbook: A Guide for Sport-Specific Performance Enhancement*, pp. 687–698. Chichester, John Wiley & Sons.

Bandura, A. (1997) *Self-Efficacy: The Exercise of Control*. W.H. Freeman, New York.

Bertram, C.P., Cote, A., Brown, M. *et al.* (2007) The effects of positive video modelling during warm-up. *Journal of Sport and Exercise Psychology* **29**, S55–S56.

Butler, R.J. and Hardy, L. (1992) The performance profile: theory and application. *The Sport Psychologist* **6**, 253–264.

Butler, R.J., Smith, M. and Irwin, I. (1993) The performance profile in practice. *Journal of Applied Sport Psychology* **5**, 48–63.

Dale, G. and Wrisberg, C. (1996) The use of a performance profile technique in a team setting: getting the athletes and coaches on the 'same page'. *The Sport Psychologist* **10**, 261–277.

Hall, C.J. and Lane, A.M. (2001) Effects of rapid weight loss on mood and performance among amateur boxers. *British Journal of Sports Medicine* **35**, 390–395.

Hatton, R. (2007) *The Hitman: My Story*. Ebury Press, London.

Ives, J.C., Straub, W.F. and Shelley, G.A. (2002) Enhancing athletic performance using digital video in consulting. *Journal of Applied Sport Psychology* **14**, 237–224.

Jones, G., Hanton, S., and Connaughton, D. (2002) What is this thing called mental toughness? An investigation of elite sport performers. *Journal of Applied Sport Psychology* **14**, 205–218.

Kelly, G.A. (1955) *The Psychology of Personal Constructs*, Vols I and II. Norton, New York.

Lane, A.M. (2006) Reflections of professional boxing consultancy. *Athletic Insight* **3**(8), 1–7. http://www.athleticinsight.com/Vol8Iss3/Reflections.htm

Lane, A.M. (2007) The rise and fall of the iceberg: development of a conceptual model of mood-performance relationships. In: Lane, A.M., (ed.), *Mood and Human Performance: Conceptual, Measurement, and Applied Issues*, pp. 1–34. Nova Science, Hauppauge, NY.

Oates, J.C. (1987) *On Boxing*. Dolphin/Doubleday, Garden City, NY.

Smith, D. and Wright, C. (2008) Imagery and sport performance. In: Lane, A.M., (ed.), *Sport and Exercise Psychology: Topics in Applied Psychology*, pp. 139–150. Hodder-Stoughton, London.

Thelwell, R. (2008) Applied sport psychology: enhancing performance using psychological skills training. In: Lane, A.M. (ed.), *Sport and Exercise Psychology: Topics in Applied Psychology*, pp. 1–15. Hodder-Stoughton, London.

Weston, N. (2008) Performance profiling. In: Lane, A.M (ed.), *Sport and Exercise Psychology: Topics in Applied Psychology*, pp. 91–108. Hodder-Stoughton, London.

5

Creating Positive Performance Beliefs: The Case of a Tenpin Bowler

Caroline Marlow

Roehampton University, London, UK

5.1 Introduction/background information

The athlete was a tenpin bowler who had represented his country at junior level and, at the time, bowled for his National Under 24 team. In the season prior to seeking psychological support, he had produced a disappointing performance at the World Youth Championships, which he considered to be due to both technical and psychological reasons. In the previous few months, he had engaged the support of a new technical coach and had come to consider his technical ability as a strength within his game. Despite these technical advances, the athlete was still not performing to his potential within competitive situations, and continued to lose to some players whom he deemed to be of a lesser ability. Consequently, he sought support to help develop a better psychological approach to his game. At the start of the consultancy period, his short-term goal was to gain selection within the following few months for that season's World Youth Championships, and his long-term goals were to be selected for his country's Senior Men's Team and to play on the European professional circuit.

Applied Sport Psychology   Edited by Brian Hemmings and Tim Holder
© 2009 John Wiley & Sons, Ltd

5.2 Initial needs assessment

The athlete's initial needs assessment was conducted through a semi-structured interview (e.g. Patton, 1990), which was held in my office and took 1.5 hours. This assessment was complimented in the second session by the implementation of an adapted version of Butler and Hardy's (1992) performance profile (as outlined later).

The semi-structured interview has become my main form of initial assessment for three reasons, each of which were reaffirmed here. First, it enables me to gain a reasonably detailed understanding of the athlete's performance background and of his/her current situation, with the obvious aim of gaining ideas for the development of an initial intervention plan. Second, I have found that performers do not always possess a great awareness of, or have lost touch with, some of the important personal aspects of their participation, or of the factors that influence their performance. Although rather neglected within the sport psychology literature, Moore and Stevenson (1994) suggested self-awareness as one of the pre-requisite characteristics for attaining the mental skills which are, in turn, essential for achieving the state of trust required for the automatic execution of performance-related skills. My interviews therefore aim to help promote this fundamental characteristic which, in turn, enables the performer to become more aware of, and engaged in, the support process. Finally, I have found semi-structured interviews to provide an important opportunity to develop a good consultant–athlete rapport. As repeatedly outlined in the counselling literature (e.g. Petitpas, Giges and Danish, 1999), the development of rapport and trust between the consultant and the athlete is one of the main determinants of successful support. Consequently, I always place much emphasis on the employment of my 'human skills', that is, actively listening to my athlete, remaining congruent and providing empathy and unconditional positive regard (Rogers, 1980).

The interview completed with the athlete comprised the following open questions, each of which were complemented by extensive use of additional elaboration and clarification probes (Patton, 1990) to more fully promote the athlete's awareness and my detailed understanding of his situation. The questions were asked in an order that was perceived to progress from the more general and easy to answer, thus promoting the opportunity for rapport development, to those that were anticipated to be more specific to attaining detailed understanding.

1. *Tell me about your bowling history.*
   This initial question allowed the athlete to discuss readily available information and thus provided an early opportunity for him to relax and for rapport to develop. It also provided useful information as to the reasons for his participation and his previous achievements/failures and critical incidents.

2. *Why are you seeking psychological support?*
   As it was the athlete's own decision to seek psychological support, this question sought to gain an insight into his perspective of his current performance situation. It also provided an opportunity to gain some awareness of his understanding of, and attitude to, potential psychological support/techniques and of his commitment to developing his mental game.

3. *What are your bowling goals?*
   This question helped to provide a context for the athlete's performance aims. Further, it also enabled me to explore the appropriateness of his current approach to goal-setting, for example, whether short/medium/long-term and outcome/performance/process goals (Burton, 1992) were appropriately operationalized.

4. *Why do you enjoy bowling?*
   Over time, I have found this to be an increasingly useful question to ask and my experiences of athletes' responses concur with Csikszentmihalyi's (1990) proposal of enjoyment being related to flow (i.e. optimal performance states). In addition, in considering this question, athletes typically change to a positive psycho-physiological state (e.g. showing positive body language, enthusiasm and empowerment), and also typically reveal those intrinsic, task-orientated reasons for participation that often prove useful for intervention development.

5. *Tell me about your best, and then typical, current performance.*
   This question specifically aimed to promote awareness and understanding of the athlete's optimal, and current sub-optimal, bowling state.

The discussion that developed from this final question led the athlete to identify that his lesser performances were most often experienced during individual matches, where he needed to perform well to gain the national selectors' attention. Conversely, his best performances typically occurred whilst bowling in team competitions. In fact, he deemed that if he bowled as well within individual matches as he did in team competitions, he would be good enough to achieve his short-term goal of being selected for the World Youth Championships. What struck me right from the start of the semi-structured interview, however, was how overt the athlete was in relaying his performance-related beliefs, and in turn, that the majority of those related to individual competition appeared to have performance-limiting potential. On perceiving this, I specifically listened out for phrases that within the cognitive therapies (e.g. Beck, 1995) are deemed indicative of underlying, irrational beliefs. These include all or nothing thinking, overgeneralizations and magnification and/or minimization. By the end of the interview, numerous beliefs were identified that, because of their occurrence or high intensity within individual, but not team, situations, appeared to have potentially performance-limiting effects. The main ones were as follows:

1. I have to be in the top four at national level U24 competitions.
2. It is about winning, and only winning matters.
3. I need to be scoring right from the start of the competition.
4. I need to score highly in every match and in every round.
5. Lesser bowlers can beat me on a 'given day'.
6. I cannot afford to make mistakes.
7. I cannot stop if I baulk (i.e. if I do not feel right prior to delivery).
8. My competitors are a danger to me, therefore I have to keep an eye on their scores.

As research highlights that elite performers generally possess a high-ego/high-task orientation profile (Hardy, Jones and Gould, 1996), my main concern was that his beliefs within the individual performance situation were predominantly ego-orientated. Further, those beliefs that did relate to his own individual performance appeared to suggest a lack of control over his behaviour and thoughts. Consequently, his beliefs generally appeared to reduce his focus on performance matters that he could control and thus what he could specifically do to attain his high performance expectations.

My questioning of how appropriate these ego-orientated beliefs were for the athlete was further confirmed by his response to the question 'Why do you enjoy bowling?' and his consideration of his team performances. First, although the athlete's immediate response as to why he enjoyed bowling was that he enjoyed winning, on greater reflection, he stated that he enjoyed the feeling of bowling well, 'knowing that you have got it just right', and the challenge of overcoming the technical difficulties of bowling: 'there's a lot to think about, every bowling lane and every ball is different, so you have to think about it'. Second, during team competitions, the athlete stated that he was not concerned about the outcome or the performance of his competitors, as he was 'more confident that we can win collectively'. I found this enhanced confidence in team matches interesting as here he appeared to focus far more on task factors (i.e. on his ability) and believed that he had not only the ability to be his team's 'winning factor', a responsibility that he appeared to relish, but also the ability to be 'the best bowler in the competition'. This latter issue was particularly pertinent as the same bowlers perform in both individual and team events.

In summary therefore, my interpretation of the interview was that, when the athlete was performing to his believed current potential (i.e. within team events), his beliefs reflected a more optimal high-ego/high-task orientation, but that when he was bowling sub-optimally (i.e. in individual competitions), his beliefs were more akin to a high-ego/low-task orientation. Reading between the lines, it appeared that the beliefs associated with each of these orientations had contrasting effects on the athlete's thoughts, emotional, physiological and behavioural reaction to events and, in turn, the performance outcome. To explain, within team matches, the athlete stated that he was confident and relaxed enough to take his time, enjoy his game, concentrate on his own performance, trust himself to make appropriate decisions and execute the appropriate bowling action smoothly. In contrast, however, within individual events he admitted that, by putting pressure on himself to win, he had created the need to be scoring highly and winning right from the start. This was fine when he started off well, but when he failed to meet his high expectations, he recounted experiencing signs of decreased self-efficacy and increased anxiety which in turn: (a) reduced his ability to read the lane correctly and thus make appropriate bowling strategy decisions; (b) made him take less time to prepare for his delivery and to continue with his delivery regardless of whether he 'felt right'; (c) made him focus on particular aspects of his bowling execution; and (d) reduced the smoothness and coordination of his bowling action. Further, regardless of how he was scoring, his beliefs regarding the need to observe his opponent's performance appeared to distract him from his own game and served to exacerbate the above consequences.

At the end of the assessment therefore, it appeared that the initial aims of any intervention programme should be: to encourage the review of, and where appropriate challenge, those beliefs that currently dominated the athlete's individual performances and to encourage the adoption of more task-orientated beliefs; to optimize, and promote the durability of, competition confidence and bowling-efficacy, and to ensure effective decision-making and automatic skill execution. With this in mind, I decided that belief challenge based on Ellis's (1973) Rationale Emotive Behaviour Therapy (REBT) should form the basis of support, complimented by the introduction of imagery into his performance preparation and current pre-performance routine. Reflective practice (Gibbs, 1988) was also encouraged throughout to promote better awareness of his performance-related beliefs, thoughts, emotions, physiology and behaviours, and the effectiveness of the intervention. The following section will explicitly outline the athlete-related and scientific rationale for the inclusion of the REBT and imagery strategies.

5.3 Interventions and monitoring

Intervention one: belief challenge

Encouraged by works such as Balague (1999) and Nesti (2004) that urge sport psychology consultants to give greater credence to an athlete's belief structure, I decided to use the second consultancy session to challenge the athlete's self-defeating, ego-orientated beliefs as a fundamental starting point for support. This decision was also made on the basis that, throughout the assessment, the athlete had demonstrated a high level of articulation and, on encouraged reflection, a good degree of personal awareness and capacity for critical thought. With this, and the high level of commitment demonstrated towards developing his psychological approach, I anticipated that the athlete would be both capable and keen to consider the appropriateness of his current performance-related belief structure.

Theoretically, my decision to start the intervention with belief challenge was based on Persons's (1989) proposal that cognitive psychology relates to a two-level model. The first, typically covert, foundational level comprises our beliefs, which have a dominating influence over our reciprocally interacting cognitions, emotions, physiology and behaviours, which form the second, more overt, level. An example of how this might help us understand this athlete's reduced performance in an individual match situation where he had not started as well as his competitors is outlined in Figure 5.1.

Dryden (2004) summarizes that, within the cognitive–behaviour-based therapies, there are two dominant forms of belief. First, irrational beliefs, those that are rigid, inconsistent with reality, illogical and self- and/or relationship-defeating, are deemed to be at the core of psychological disturbance. Second, rational beliefs, that are flexible, consistent with reality, logical and self- and/or relationship-enhancing, are at the core of psychological health. The aim of this first intervention process, therefore, was to help the athlete consider the potentially self-defeating aspects of his current performance-related belief structure and, where appropriate, to change these beliefs to those that are more self-enhancing.

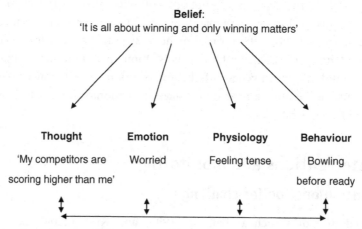

Figure 5.1 An example of the proposed relationship between beliefs and thoughts, emotions, physiology and behaviour.

Ellis's (1973) process of REBT, acclaimed as the foundation of the modern cognitive–behaviour therapies (Beck, 2007), was deemed appropriate to address and challenge each of these beliefs in turn. Historically, REBT has been employed with those displaying neurotic behaviour; however, the use of this technique with non-clinical populations has become increasingly common. For example, Ellis himself (1994) advocated REBT for the promotion of exercise behaviour in sedentary populations, whilst Criddle (2007) and Trip, Vernon and McMahon (2007) have more recently highlighted the success of its application to business coaching and education, respectively. Despite this, REBT has attained little attention within the sport psychology literature, so consequently, the basic premises of REBT will be outlined here.

REBT originally stemmed from the humanistic approach (Ellis, 1973), whereby individuals are viewed as holistic, creative beings, with the ability to direct their own destinies. The focus therefore is on the exploration of the individual's basic experience and values with the aim of promoting and fulfilling the individual's right to achieve self-actualization. Despite the acceptance of human independence and of the individual's ability to choose his or her way of achieving fulfilment, many of us, as Ellis (1982) suggested, inadvertently sabotage our efforts to optimize our happiness and the attainment of goals by the adoption of irrational beliefs. REBT therefore aims to assist individuals in the identification, challenge and replacement of such self-defeating beliefs.

REBT follows an ABC(DE) model where, as an individual encounters an *Activating* event (A), his or her *Beliefs* (B) will determine the emotional or behavioural *Consequence* (C). As implied above, Irrational Beliefs (iBs) lead to self-defeating consequences whereas Rational Beliefs (rBs) promote more fulfilling outcomes. My task, therefore, was to help the athlete to consider his performance experiences with the intention of moving him towards being able to *Dispute* (D) his irrational beliefs, with the final aim of enabling him to develop an *Effective* new philosophy (E) that would better support his performance aims. The new philosophy therefore should provide rational beliefs that promote the consequence of attaining his optimal performance state when faced with particular activating events, such as national-level, individual competitions. The following process, adapted from Ellis' (1994) REBT, was therefore conducted throughout the second session.

To start the intervention, I relayed to the athlete my perception that he held different beliefs in individual and team performances. Then, using a diagram such as in Figure 5.1, I explained how beliefs could influence performance-related thoughts, emotions, physiology and behaviour, and thus how I perceived that his beliefs might be leading to his different levels of performance. As this gained the athlete's interest, I continued to explain how REBT's ABC(DE) model could help him consider and change his beliefs as appropriate. The athlete agreed that this was a potentially appropriate form of intervention. To continue, it was first important to ascertain that the beliefs that I had identified throughout the interview were indeed held by the athlete. To do this, and to visually outline the differences that I expected to typify his team and individual competitive performances, I decided to utilize the framework, and an adapted form, of the performance profile process (Butler and Hardy, 1992). This process is outlined below.

1. The athlete was guided to reflect on a specific, recent, but typical example of his individual bowling performance.

2. The athlete was asked to consider the beliefs, thoughts, emotions, physiology and behaviours that he experienced in this individual performance.

3. These elements were added onto the bar chart template of the performance profile.

4. Time was taken to ensure that the identified elements were indeed representative of the athlete's experience and that any further elements (particularly those evident from the assessment interview) were added as appropriate.

5. The athlete was encouraged to explore how each of these elements contributed to his identified *individual* performance, which culminated in him rating the intensity of each belief and of the identified thoughts, emotions and physiology on a scale of 0–10 (where 0 = not present, and 10 = extremely intense), and also to his rating of the frequency of behaviours (where 0 = not present, and 10 = extremely frequent).

6. Steps 1–5 above were then repeated, this time considering a specific, recent and typical example of his *team* bowling performance.

7. To explicitly highlight the differences between the two situations, the elements and ratings for the *team* performance were added to the previously compiled *individual* profile so that it contained elements associated with both performance situations.

8. To complete the comparison, any element that had been identified in one performance situation but not the other was considered in the situation where it had not previously been considered so that eventually all elements were rated for both team and individual performance situations.

For clarity here, the resulting profile is presented in two sections. Figure 5.2 outlines the intensity of the beliefs held on a scale of 0 (not present) to 10 (extremely intense) for both his typical, current team and individual performance examples. Figure 5.3 outlines how the athlete categorized elements into performance-related thoughts, emotions, physiology and behaviours and again how he associated them with the team

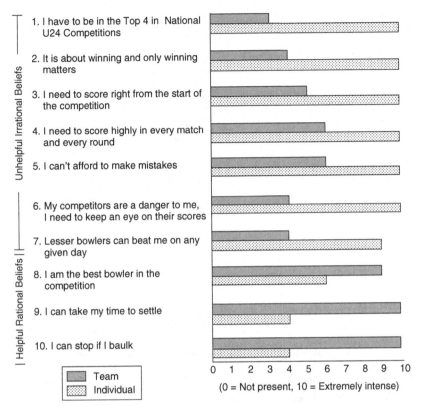

Figure 5.2 Pre-intervention profile of the individual and team performance belief intensities.

and individual performance situations. Different reference points were used for the rating of these, with thoughts, emotions and physiology being considered in terms of their intensity and behaviours in terms of their frequency. Both are scored on a scale of 0 (not present) to 10 (extremely intense or frequent).

The profile very effectively highlighted the ABC component of the REBT model (Ellis, 1994), as the athlete quickly came to realize either that he held different beliefs for team and individual competitions or that the beliefs held were at different intensities in the two situations. Further, he came to associate the held beliefs with his relative performance thoughts, emotions, physiology and behaviours, and indeed the performance outcome. Specifically, he realized that the first seven beliefs on Figure 5.2 appeared to relate to his sub-optimal performance in *individual* matches (i.e. were unhelpful and irrational beliefs, as indicated by the low individual ratings on all the elements in Figure 5.3), whilst the remaining three beliefs appeared to lead to his optimal performance within the team situations (i.e. were helpful and rational beliefs as indicated by the high team ratings on all the elements within Figure 5.3). On

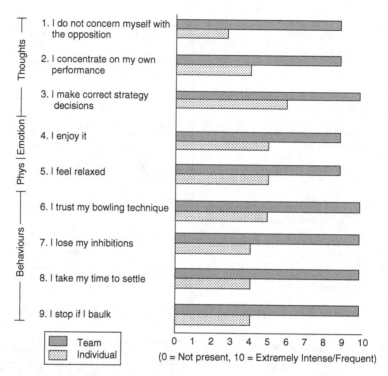

Figure 5.3 Pre-intervention profile of individual and team performance thoughts, emotion, physiology and behaviour.

appreciating this, the athlete was keen to start changing his beliefs and identified 'It is all about winning. Only winning matters' as the irrational belief most instrumental to reducing his individual performance. To continue through the ABC(DE) model, therefore, I encouraged the athlete to *Dispute* this belief by carefully considering from his experience how this belief had, or had not, helped his performance. Here he realized that any national level, individual competition had become an *Activating* Event that led to this *Belief* with the *Consequence* that he came to pay too much attention to his competitors as he looked to see how he was performing relative to them, and put too much pressure on himself to start scoring right from the beginning of the match. Previously he had perceived that these thoughts and behaviours gave him an advantage over his competitors, who were wary of him, but now on reflection he realized that knowing how his opponents were scoring, when combined with his initial high scoring expectations could easily lead to rapid losses in confidence, which in turn reduced his ability to lose his inhibitions, relax and start again/compose himself if he was not feeling ready to perform.

With the belief disputed, we started the final REBT stage, to consider a more *Effective* New Philosophy (i.e. a rational, alternative belief that would enable the athlete to concentrate more fully on his individual performance). Again here, the completion of the adapted performance profile proved highly beneficial as it provided potential rational beliefs that the athlete already held within team competitions and, therefore, evidence that they were associated with optimal performance. Consequently, he readily reviewed their suitability to the individual competitive environment. The athlete introduced a new belief, based on previously identified thoughts, that 'I can allow myself to enjoy bowling'. An overview of the ABC(DE) process completed is outlined in Figure 5.4.

To conclude, it was important to double check the ecology of the new belief (i.e. to ensure that the belief was appropriate to the individual performance situation). Again the testing of this belief was carried out in accordance with REBT recommended practice (Ellis and Dryden, 1987). First, the athlete was encouraged to gain a real embodied sense of what the 'I can allow myself to enjoy bowling' belief might feel like, and then to visually rehearse holding this belief through a competitive performance. Within this, the athlete was encouraged to notice how this new belief influenced his thoughts, emotions, physiology and behaviours. On reflection of this experience, the athlete noted, 'I love the feel of knowing that I've got it right and just watching the ball go down the lane', 'I like the technical elements, trying to bowl the perfect ball' and 'I enjoy the tactical elements. You really have to think about how to play each

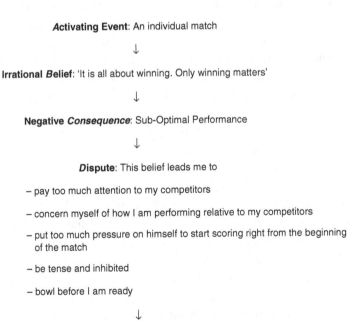

Activating Event: An individual match

↓

Irrational *Belief*: 'It is all about winning. Only winning matters'

↓

Negative *Consequence*: Sub-Optimal Performance

↓

Dispute: This belief leads me to

– pay too much attention to my competitors

– concern myself of how I am performing relative to my competitors

– put too much pressure on himself to start scoring right from the beginning of the match

– be tense and inhibited

– bowl before I am ready

↓

***Effective* New Philosophy**: 'I can allow myself to enjoy bowling'

Figure 5.4 An overview of the ABC(DE) process completed by the athlete.

ball. No game/day is ever the same'. These further promoted my confidence in the new belief as they were similar to those answers provided to the 'Why do you enjoy bowling?' question within the assessment interview. At the end of the session, the athlete was encouraged to continue to check the ecology of the belief by going away and simulating a competitive situation within training and seeing whether it was possible to hold the belief. Further, as a major strategy of REBT is to help the athlete to attain autonomy and to take responsibility for their own change processes (Ellis, 1994), the athlete was also encouraged to engage in reflective practice (Gibbs, 1988) to identify any further potentially limiting beliefs. The third consultancy session therefore enabled the athlete to report that he had fully developed the belief, 'If I focus on enjoying myself and performing to the best of my ability, the competition will take care of itself' and that he thought it was already improving his individual match performance. The remainder of the third session was used to continue to reflect on the benefits of this change and to ensure that the athlete would know how to use the ABC(DE) model in future.

Intervention two: imagery

Although Persons's (1989) two-level model of cognitive psychology proposed beliefs as the dominating influence over thoughts, emotions, physiology and behaviours, it also recognizes that this second level can have a reciprocal effect on beliefs. In the fourth consultancy session, I therefore outlined the four basic psychological skills (Hardy, Jones and Gould, 1996) that aim to intervene at this level and asked the athlete which he thought most appropriate for supporting his belief-change intervention. Having heard of bowlers on the American professional circuit employing imagery, he was keen to explore its potential despite having no personal experience of using it or real knowledge of its potential uses. The first steps, therefore, were to inform him of the basic skill and uses of imagery, and then to develop and apply these imagery skills to support more appropriate belief structures. This process was completed in three imagery-based sessions as outlined below.

Gregg and Hall (2005) suggested that athletes are more likely to use the imagery techniques they are comfortable with, whilst Vadocz, Hall and Moritz (1997) demonstrated a cyclical relationship between imagery ability and imagery use. Here, an enhanced ability encourages more frequent use, which consequently leads to gains in imagery skill. Further, Martin, Moritz and Hall (1999) suggested that imagery ability may moderate the link between imagery use and outcome. It was therefore deemed important to spend time developing the athlete's imagery skills to ensure that he was comfortable using them. More specifically, attention was paid to enhancing visual and kinaesthetic imagery ability, as evidence suggests that both of these modes are related to motivational general-arousal (MG-A) and motivational general-mastery (MG-M) imagery use (see later in this section for an explanation of these terms) (Moritz, Hall, Martin, and Vadocz, 1996; Vadocz *et al.*, 1997), which, on initial consideration, were two of the functions deemed to be important for this athlete.

I have used imagery extensively across athletes of different sports, ages and competitive levels and, in recent years, have increasingly employed those imagery techniques originally noted by Bandler (1985), with seemingly excellent results in terms of finding an approach to imagery that an athlete enjoys using and in optimizing the athlete's experience. Using this neuro-linguistic programming (NLP) approach, the athlete was encouraged within the first imagery session to explore the psycho-physiological effects of changing the sub-modalities of an image. To explain, individuals access information through either visual, kinaesthetic or auditory processes or modalities. The experience of using these sensory modalities can be broken down into component parts or sub-modalities; for example, visual images can be viewed in terms of their size, brightness or colours, auditory imagery can be heard in relation to the volume, pitch and tone, and kinaesthetic images can be felt in terms of the intensity and temperature. In accordance with the sensory modality that the individual prefers to use to access information, therefore, sub-modality changes, such as those outlined above, have an effect on the individual's psycho-physiological state which, in turn, enhances the imagery experience (Bandler and MacDonald, 1988). Consequently, I asked the athlete to try and imagine a recent bowling experience and then guided him to experiment with his image by changing the various visual, auditory and kinaesthetic sub-modalities until a rich experience was attained. The athlete reported that three sub-modality changes in particular produced the most relevant and detailed visual and kinaesthetic experience. First, the introduction of bolder, brighter colours for the important aspects of his performance picture gave him greater clarity for strategic decision-making and then confidence in the strategy's execution. Specifically, he imagined the ball, lane and desired bowling line as a vivid blue, red and gold respectively (whilst other aspects of the picture dulled and faded). Second, he reported that by increasing his body size relative to the surrounding environment he experienced a substantial increase in his feelings of confidence. Finally, as the combination of the previous two images enabled the athlete to gain a kinaesthetic experience of his optimal bowling state, he identified the essential presence of a warm sensation in his finger tips which he came to term 'confident precision'. By further extending this sensation from his finger tips up into the shoulder of his right (bowling) arm, the athlete experienced a greater sense of technical control. Although not always easy, overall the athlete found this imagery exploration to be a powerfully positive, but surreal experience, and following discussion of further, potential sub-modality changes, he was keen to experiment with them further to enable him to become more familiar with the process and its possibilities. The athlete was encouraged to do so within a relaxed environment for 15 minutes on a regular basis (i.e. daily) throughout the following week.

By the fifth session, the athlete had become more familiar with using imagery, so my aim here was to help him more fully experience its benefits. As previously mentioned, when designing imagery interventions, Gregg and Hall (2005) encouraged matching the type (function) of imagery employed with the task demands to ensure that the most effective outcome is achieved. Hall, Mack, Paivio, and Hausenblas (1998) identified five functions of imagery use in sport. From the needs analysis, the main purpose for the implementation of imagery was to rehearse and to interject the newly developed beliefs into the athlete's individual performance, with the desired outcome of improving the athlete's confidence both generally throughout a competition and more specifically in his strategic decisions. In this session, the athlete was informed of, and guided through, imagery sessions relating to MG-A and MG-M imagery as outlined below.

First, MG-A imagery, which involves the creation of images that include the arousal and anxiety associated with performing, was encouraged. Hecker and Kaczor (1998) showed that such imagery is capable of elevating baseline heart rates to that experienced in an actual situation, which has consequently been found to decrease anxiety (Vadocz et al., 1997) and to be related to an increase in confidence (e.g. Abma, Fry, Li, and Relyea, 2002; Mills, Munroe and Hall, 2000). It was intended that through the experience of MG-A imagery the athlete could consider previous situations where high levels of arousal and anxiety were experienced and then come to control the situation through the use of his new more appropriate belief structures. Specifically, I guided him through an imagery scenario where he was concerned by his opponents scoring higher than him and consequently feeling tense and inhibited. This, in turn, led him to bowl before he was ready and to make bad tactical decisions. To counteract this, I encouraged him to imagine bowling whilst focusing on the belief, 'If I focus on enjoying myself and performing to the best of my ability, the competition will take care of itself'. Then to reinforce the effectiveness of this belief, I encouraged him to employ the previously highlighted positive sub-modality changes (i.e. increasing his body size relative to his surrounding and increasing the colours and brightness of relevant surroundings, whilst dimming non-relevant surroundings). Finally, I encouraged him to spend time fully reinforcing the association between this new rational and task-oriented belief and thus the positive changes to his confidence and ability to focus on his own performance, and consequently his improved decision-making and ability to relax/lose his inhibitions.

Second, the athlete was guided through an example of MG-M imagery, which involves imaging performance whilst remaining in control and feeling confident. Short, Smiley and Ross-Stewart's (2005) review stated that the relationship between MG-M imagery use and self-efficacy is robust across studies. It was therefore intended that the use of this type of imagery would help the athlete to prepare to start with confidence and effective rational beliefs in mind, and then to maintain it throughout each match and the competition as a whole. Here, I guided the athlete to consider starting an individual competition with the beliefs 'I am the best bowler in the competition' and 'I can stay in control, take my time, and allow myself to enjoy bowling'. Again by working on the sub-modalities (i.e. bringing the athlete's attention to, and then expanding, his feeling of 'confident precision', and by him seeing the ball, lane and desired bowling line in vivid colours), I helped the athlete attain a highly optimal performance state. Again I reinforced the association between this optimal performance state and the rational, task-oriented belief. At the end of this session, the athlete was encouraged to complete these two forms of imagery on a regular basis, but especially the night before, and the morning of, a competition, to promote the likelihood of starting the match confident.

At the start of the final imagery session, the athlete reported enjoying using the MG-A and MG-M imagery and that he had become much better able to maintain more effective beliefs and to remain confident during individual performances. To further ensure the robustness of the athlete's new belief structure and to maintain focus on his technical skills, I decided to introduce the third of Hall *et al.*'s (1998) imagery functions, cognitive general (CG) imagery. This involves the mental rehearsal of competition-related strategy, and has been found by Abma *et al.* (2002) and Callow and Hardy (2001) to have a positive relationship with confidence in track and field athletes and netballers respectively. The aim here was to introduce CG imagery into his existing behavioural pre-performance routine (PPR) to help reinforce his bowling strategy decisions and to instigate the required movement pattern prior to the initiation of the bowling action. Specifically, Radlo, Steinberg, Singer, Barba and Melnikov (2002) reported that focusing on the target (i.e. the use of an external, attentional focus strategy) reduced electro-cortical activity and heart rate, and had greater performance-enhancing effects than a self-directed strategy where attention is focused on the body and physical execution of a skill. The athlete, who reported already looking down the lane at the pins prior to bowling, was therefore encouraged to introduce imaging the 'vivid blue' ball, rolling along the 'vivid gold' desired path and scoring a 'strike' (i.e. employing helpful sub-modality changes again). To reinforce this, I guided him to image employing this imagery immediately prior to ball delivery, ensuring that he was using strong colours (as outlined above) to emphasize the delivery's success.

Once the athlete was comfortable with the idea of introducing CG imagery into his pre-performance routine, we continued to discuss how he might better use his time immediately after and in-between balls. Here, he was encouraged to relax by either fully engaging with those factors that promote his enjoyment, or by completing either the previously practised MG-M or MG-A imagery to help him focus on appropriate beliefs. Again, here I guided him through imaging an entire bowling cycle using the MG-M and MG-A imagery in-between balls, and CG imagery immediately prior to delivery, and using the sub-modalities to maximize the experience of the optimal bowling state.

As with other skills, a change to a PPR needs to be practised in a training environment prior to use within competition so that it can become an automatic component of the performance. The athlete was therefore encouraged to rehearse his new PPR in training and then to slowly introduce it to increasing levels of competition as he felt appropriate.

5.4 Evaluation of intervention

The monitoring of an intervention's success and of my own personal delivery style is fundamental to my consultancy work. Consequently, evaluation is treated as a constant, on-going process. Because of the amount and depth of information that can be attained, the majority of my evaluations are conducted qualitatively in both a formal and an informal manner. Throughout each session, I continually asked for the athlete's perception of an intervention's efficacy, to ensure that the high efficacy required to promote adherence was held. Further, at the end of each session, I asked for his overall thoughts regarding the content and delivery style. This provided an extra opportunity to check the athlete's awareness and perceptions of the process experienced and therefore, if appropriate, for further contributions or clarifications to be made. In-between sessions, the athlete was keen to e-mail feedback of his competitive performance and future process goals, which enabled further, informal advice and support to be provided. Finally, at the start of each consultation, the athlete provided further feedback on his ability to implement the interventions and their effect on his training/competitive performance. Recommendations on how to adapt the intervention or its implementation to better suit the athlete's practical and personal preferences were therefore considered as appropriate.

Throughout the consultancy process, the athlete was very supportive of our work. The following e-mail extracts highlight the developments in his game and how he felt that they led to the achievement of his short-term goal, to gain selection for the World Youth Championships. Following a qualifying tournament for the World Youth Championships (the first after the start of our consultancy support),

> I finished second today. I played really well and was really focused. I eventually lost out by 1 pin. Rather annoying, but I performed great, so I'm happy. People commented today that I'm looking a lot more focused and controlled than ever, so we are definitely getting somewhere!

Following a club-level game,

> I don't want to seem like I keep going on and on, but I had another good day today. I smashed the all-time individual scoring record by over 100 pins! Although it was only a small

tournament, that's probably the best I have ever played. I posted an average of 267!! It's also my personal best 3 game score. The best part about it all, I feel, was that throughout I had belief that every shot was going to be a strike (30 out of 34 balls were strikes). Fingers crossed I can keep my form for next week. Absolutely no reason why I can't win.

Following a further qualifying tournament for the World Youth Championships,

I'm going to the World Youth Championships! Well things worked out for me on Sunday. I bowled well and won the tournament, securing my place in [his national] team. I'm especially glad because it wasn't one of those days where everything goes right for me. In fact it was quite tough going, I didn't get a whole lot of luck. But I'm very happy, I managed to come through and get the job done so I'm on my way to meeting my goals this season.

This last e-mail was particularly informative as it demonstrated that the athlete had managed to employ his belief and imagery strategies to help promote his optimal bowling state even when the competition was going against him. Although the athlete's results highlighted a substantial improvement in his individual performance, it was appropriate to end our consultancy period by more formally assessing the effectiveness of the intervention. Consequently, the beliefs, thoughts, emotions, physiology and behaviours considered within the initial performance profiles were revisited to see whether any specific changes had occurred. Figure 5.5 shows the resultant changes in the intensity of the athlete's beliefs with the bar lines showing the pre-intervention score and the arrows showing the direction and amount of change in belief intensity. Here it became apparent that although, the intensity of the belief 'I have to remain in the Top 4 in National U24 competitions' remained high, the intensity of the other ego-oriented, irrational beliefs (i.e. Figure 5.5, beliefs 2–7) that he deemed to be unhelpful had all reduced. Further, along with the addition of the two helpful, rational beliefs that were developed as a result of the intervention (i.e. belief 8, 'I can allow myself to enjoy bowling' and belief 9, 'I can focus on performing to the best of my ability'), the intensity of the three initial rational beliefs (10–12) that enabled him to maintain control of the important task-orientated elements of his performance also increased within his individual game. Overall, therefore, those ego-orientated beliefs that appeared to limit his performance reduced in intensity whilst those task-orientated beliefs that promoted performance increased to produce a profile far more akin to that initially associated with his team performance. Figure 5.6 further supports the positive nature of these belief changes. Here the arrows highlight that the athlete had become better at establishing and maintaining performance-enhancing thoughts, emotions, physiology and behaviours within his individual performance, so that again they were more akin to his team performances.

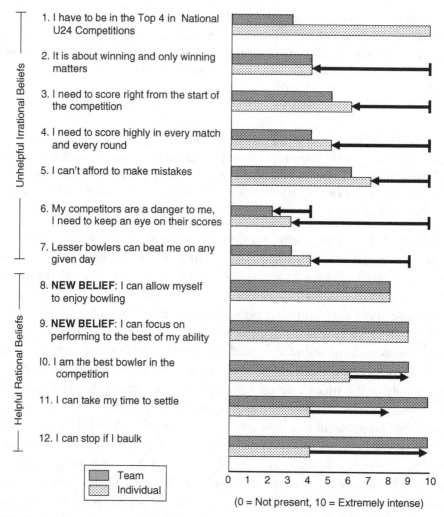

Figure 5.5   Pre- to post-intervention profile showing changes in individual and team performance belief intensities.

5.5 Evaluation of consultant effectiveness/reflective practice

From conversations that I have had with other sport psychologists I have come to realize that I am quite creative in the support that I offer my athletes. I think that this lies in my approach where I am determined to provide the best possible form of support, and probably the confidence in my practical and personal skills to indulge my own intellectual curiosity to investigate new approaches. In the past, I have often experienced a concern where, despite my confidence in my ability to assist an athlete, I am aware

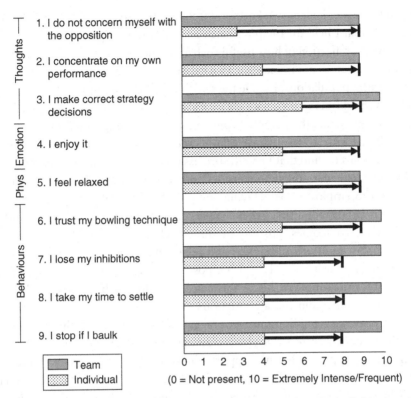

Figure 5.6 Pre- to post-intervention profile showing changes in individual and team performance thoughts, emotion, physiology and behaviour.

that my methods are not always evidenced within the sport psychology literature, and therefore open to criticism from others within the profession. On deeper reflection now, however, I believe that, to the contrary, my constant engagement in literature from across the psychology-related disciplines and from varying philosophies has given me the assurance that (a) my support work is justified within the broader psychology field, and (b) I have a fuller understanding of the boundaries that are appropriate or not appropriate for me to cross. I feel that this paper has provided several examples of how, even though I have used methods and techniques not common within the sport psychology literature, I have stayed within the boundaries of our professional codes of conduct. Two examples shall be reflected upon here.

First, although on occasion I have worked with athletes to change their beliefs, this was the first time that belief change formed the primary aim of support, that I was so explicit in this aim with the athlete and that I used a specific belief change strategy (i.e. REBT). My previous concern at using belief change was probably caused by the dominant focus within the literature of intervention at the second, lower level (i.e. to physiology, thoughts, emotions and behaviours) and consequently a lack of reference to change at the higher belief level. Further, I feel this was also possibly due to the 'clinical associations' that I held for methods, such as REBT, that originate from mainstream psychology. On reflection of how useful such a technique might have been in previous client work, however, and in realizing that belief change can be equally as empowering for non-clinical populations, I made the decision that the use of REBT was within my boundaries should the appropriate client and situation materialize. As described, the assessment highlighted the potential appropriateness of using REBT with this athlete. This, combined with my confidence that his was an entirely performance-related and non-clinical concern, and that I had the professional expertise and empathetic capacity to guide this particular athlete to consider and challenge his beliefs, helped me make the decision to employ REBT. Having completed the REBT process, I can support Persons's (1989) proposal that belief change can have a powerful and rapid influence on performance outcomes, possibly by directly changing the athlete's performance related thoughts, emotions, physiology and behaviours. In addition, due to the athlete's desire to continue using REBT, and the positive changes that he experienced in his enjoyment of, and attitude to, his performance, I would now be more confident to use REBT should another appropriate situation arise.

Second, I have been guiding and encouraging athletes to play with the sub-modalities within their imagery since I completed training in such techniques a couple of years ago. Again, this approach has no support within the sport psychology literature, but through my own personal trials and by guiding colleagues to change the sub-modalities in their images, I have become assured that this technique had potential benefits for client athletes. I was enthused at the success of this approach at changing the bowler's sub-modalities; this helped to improve his imagery ability, brought about powerful second level responses (i.e. to his imagined thoughts, emotions and physiology) and provided a technique that he was keen to experiment with and employ.

To end my reflection, I will answer a question that I have often been asked; 'What is the most important attribute of a good sport psychologist?' From going through the process of writing this case study, I have come to realize that my answer is now 'informed flexibility'. My ability to provide effective support for an athlete is based on the flexibility that enables me to discover the real crux of an athlete's reason for coming to seek assistance. Once this has been achieved, flexibility is required to complete the on-going process of finding the most appropriate way of supporting that specific athlete. I now believe, however, that I only have the confidence to provide this required flexibility due to my constant engagement with the broader literature and my reflections on my consultancy experiences and skill base. These reflections have therefore provided me with the 'informed flexibility' required to provide the most appropriate support that I can to my athletes, whilst ensuring that I remain within the boundaries of my expertise. In a cyclical fashion, it is my ability to do this that continues to fuel my passion for our profession.

5.6 Summary

Currently, sport psychology interventions typically aim to promote an athlete's performance by changing his or her performance-related thoughts, emotions, physiology and/or behaviours. Persons's (1989) two-level model of cognitive psychology proposes that these are in fact strongly influenced by the individual's beliefs, with rational beliefs promoting, and irrational beliefs reducing, performance. Consequently, this chapter has described how the principles of REBT (Ellis, 1994), a common technique within the cognitive therapies, were implemented to directly influence the performance-related beliefs of a national level junior tenpin bowler.

The initial needs analysis highlighted that the athlete performed better in team, than in individual, competitions. Further, in individual competitions the athlete had a far stronger ego-orientated profile than in team situations. Using an adapted form of the performance profile (Butler and Hardy, 1992), the athlete was guided through the five stages of REBT. This entailed coming to understand how the beliefs that he held in each performance situation influenced his performance, and then challenging irrational beliefs so that they could be replaced with more helpful rational beliefs. The athlete was then taught how to use imagery to practise holding his new beliefs within individual performance situations whilst away from competition, and as part of a pre-performance routine to help maintain rational beliefs whilst competing.

Questions for students

1 Identify a number of beliefs that reflect first, an ego-orientation, and second, a task-orientation.

2 Identify a number of rational and irrational beliefs that might influence a sport performer.

3 How could the REBT's ABC(DE) model highlighted in this chapter be used to challenge and replace the irrational beliefs identified in question 2?

4 Devise an imagery scenario that would help an athlete mentally practice one of the new rational beliefs identified in question 3.

5 What athlete characteristics would be required for a sport psychologist to consider utilizing a belief change intervention?

The athlete's and my own reflections of the consultancy process, and the athlete's improved performance within individual competitions, highlights the effectiveness of the intervention. Fellow sport psychologists are encouraged to constantly review and expand their support techniques to ensure that the individual demands of athletes are met.

References

Abma, C., Fry., M., Li, Y. *et al.* (2002) Differences in imagery content and imagery ability between high and low confident track and field athletes. *Journal of Applied Sport Psychology* **14**, 67–75.

Balague, G. (1999) Understanding identity, value and meaning when working with elite athletes. *The Sport Psychologist* **13**, 89–98.

Bandler, R. (1985) *Using Your Brain for a Change: Neuro-Linguistic Programming*. Real People Press, Moab, UT.

Bandler, R. and MacDonald, W. (1988) *An Insiders Guide to Sub-modalities*. Meta Publications, Capitola, CA.

Beck, A.T. (2007) Editorial. *Journal of Cognitive and Behavioural Psychotherapies* **7**, 125–126.

Beck, J.S. (1995) *Cognitive Therapy: Basics and Beyond*. Guilford, New York.

Burton, D. (1992) The Jekyll and Hyde nature of goals: reconceptualising goal-setting in sport. In: Horn, T. (ed.), *Advances in Sport Psychology*, pp. 267–297. Human Kinetics, Champaign, IL.

Butler, R.J. and Hardy, L. (1992) The performance profile: theory and application. *The Sport Psychologist* **6**, 253–264.

Callow, N. and Hardy, L. (2001) Types of imagery associated with sport confidence in netball players of varying skill levels. *Journal of Applied Sport Psychology* **13**, 1–17.

Criddle, W.D. (2007) Adapting REBT to the world of business. *Journal of Rational-Emotive and Cognitive Behaviour Therapy* **25**, 87–106.

Csikszentmihalyi, M. (1990) *Flow: The Psychology of Optimal Experience*. Harper and Row, New York.

Cumming, J. and Hall, C. (2002) Athletes' use of imagery in the off-season. *The Sport Psychologist* **16**, 160–172.

Dryden, W. (2004) Rationale emotive behaviour therapy. In: Feltham, C. and Horton, I. (eds), *Handbook of Counselling and Psychotherapy*, pp. 326–330 Sage, London.

Ellis, A. (1973) *Humanistic Psychotherapy: the Rationale-Emotive Approach*. Crown and McGraw-Hill, New York.

Ellis, A. (1982) Self-direction in sport and life. In: Orlick, T., Partington, J.T and J.H. Salmela (eds), *Mental Training for Coaches and Athletes*, pp. 10–18. Coaching Association of Canada, Ottawa.

Ellis, A. (1994) Rational emotive behaviour therapy: an application to exercise avoidance. *The Sport Psychologist* **8**, 247–261.

Ellis, A. and Dryden, W. (1987) *The Practice of Rationale Emotive Therapy*. Springer, New York.

Gibbs, G. (1988) *Learning by Doing: A Guide to Teaching and Learning Methods*. Further Education Unit, London.

Gregg, M. and Hall, C. (2005) The imagery ability, imagery use, and performance relationship. *The Sport Psychologist* **19**, 9–99.

Hall, C.R., Mack, D.E., Paivio, A. *et al*. (1998) Imagery use by athletes: development of the sport imagery questionnaire. *International Journal of Sport Psychology* **29**, 73–89.

Hardy, L., Jones G. and Gould, D. (1996) *Understanding Psychological Preparation for Sport: Theory and Practice of Elite Performers*. Wiley, Chichester.

Hecker, J.E. and Kaczor, L.M. (1998) Application of imagery theory to sport psychology: some preliminary findings. *Journal of Sport and Exercise Psychology* **10**, 363–373.

Martin, K.A., Moritz, S.E. and Hall, C.R. (1999) Imagery use in sport: a literature review and applied model. *The Sport Psychologist* **13**, 245–268.

Mills, K.D., Munroe, K.J. and Hall, C.R. (2000) The relationship between imagery and self-efficacy in competitive athletics. *Imagination, Cognition and Personality* **20**, 33–39.

Moore, W.E. and Stevenson, J.R. (1994) Training for trust in sports skills. *The Sport Psychologist* **8**, 1–12.

Moritz, S.E., Hall, C.R., Martin, K.A. *et al*. (1996) What are confident athletes imaging? An examination of image content. *The Sport Psychologist* **10**, 171–179.

Nesti, M. (2004) *Existential Psychology and Sport*. Routledge, Abdingdon.

Patton, M.Q. (1990) *Qualitative Evaluation and Research Methods*. Sage, Newbury Park, CA.

Persons, J.B. (1989) *Cognitive Therapy in Practice: A Case Formulation Approach*. Norton, New York.

Petitpas, A.J., Giges, B. and Danish, S.J. (1999) The sport psychologist–athlete relationship: implications for training. *The Sport Psychologist* **13**, 344–357.

Radlo, S.J., Steinberg, G.M., Singer, R.N. *et al*. (2002) The influence of an attentional focus strategy on alpha brain wave activity, heart rate, and dart-throwing performance. *International Journal of Sport Psychology* **33**, 205–217.

Rogers, C. (1980) *A Way of Being*. Houghton Mifflin, Boston, MA.

Short, S.E., Smiley, M. and Ross-Stewart, L. (2005) The relationship between efficacy beliefs and imagery use in coaches. *The Sport Psychologist* **19**, 380–394.

Singer, R.N. (1988) Strategies and metastrategies in learning and performing self-paced athletic skills. *The Sport Psychologist* **2**, 49–68.

Trip, S., Vernon, A. and McMahon, J. (2007) Effectiveness of rational-emotive education: a quantitative meta-analytical study. *Journal of Cognitive and Behavioral Psychotherapies* **7**, 81–93.

Vadocz, E.A., Hall, C.R. and Moritz, S.E. (1997) The relationship between competitive anxiety and imagery use. *Journal of Applied Sport Psychology* **9**, 241–253.

6
Enhancing Confidence in a Youth Golfer

Iain Greenlees
University of Chichester, Chichester, UK

6.1 Introduction/background information

The support work that will be described in this chapter involves a junior golfer (hereafter named Christopher) with low levels of confidence. Chris's parents made contact with me through the County Junior Golf Organizer, whom I had worked with during English Golf Union (EGU) training schemes for promising young golfers. Over the telephone they informed me that their son was 15 years old and a promising county-level golfer (handicap of 4). Chris's parents suggested that their reasons for getting in touch were threefold. First, Chris and his parents felt that he should be more confident than he actually was. Second, Chris's rate of progress in reducing his playing handicap had slowed and he was getting frustrated at this and felt that sport psychology might be a way of improving. Finally, the family also wanted to learn more about how sport psychology could help Chris's enjoyment of golf.

This case study represents the first stage in my work with Chris, and details the first eight meetings we had. This covered a period of four months. The consultancy with the individual has now finished, although I still occasionally receive an e-mail or phone call from Chris to discuss the psychology of golf. I have presented the first part of the intervention only, as the meetings represent a distinct intervention with

Applied Sport Psychology   Edited by Brian Hemmings and Tim Holder
© 2009 John Wiley & Sons, Ltd

this particular golfer (psychological preparation for the golfing season with confidence levels specifically targeted).

6.2 Initial needs assessment

In general, my aims for any initial assessment session are twofold. First, and fundamentally, my aim is to establish rapport, trust and credibility with the athlete. The quality of the practitioner–client relationship has frequently been identified (e.g. Ravizza, 1988) as essential for a successful consultancy. The second aim is to get a detailed understanding of the athlete's experience of their sport and their psychological approach to their sport. The initial assessment and preliminary discussion of potential mental skill interventions with Christopher were scheduled for a 2 hour session. This length of time is what I would normally set aside for a first session in these circumstances. This is purely a personal preference based on my own experiences and reflections on previous consultations with junior golfers. One hour or less gives insufficient time to develop a full understanding of the client and I have felt that the sessions have been overly rushed. However, for anything longer than 2 hours I have found it difficult to maintain the required levels of focus as a sport psychologist, and young clients have also struggled. Chris came with both his parents, who were invited (by Chris and myself) to sit in on the session on the proviso that they allowed Chris to respond fully to all questions and to have input only when invited. The session was conducted in a quiet common room at my university. This was chosen to allow the parents to be present in the room during the interview but, if requested by the performer, to be removed from the discussion.

The first part of the initial meeting with Chris was a discussion of sport psychology, the services provided by sport psychologists in general and myself in particular. This has been proposed (e.g. Ravizza, 1988; Salacuse, 1994; Pocwardowski, Sherman and Henschen, 1998) as an important activity at the onset of a consultancy as it allows for roles to be clarified and working practices to be clear from the start of the working relationship. In the discussion, I was interested to find out what Chris's expectations were concerning working with a sport psychologist, to rectify any worries or misconceptions that he may have (in this case there were none but it is not uncommon for clients to believe that sport psychology represents a 'quick fix' or that they will have a relatively passive role in the relationship), and to hear any concerns he had about what we would be doing together. This part of the meeting also gave me a chance to explain my working practices and what I expected from Chris in terms of time commitment if he chose to continue to work with me. It also gave me an opportunity to provide information from BASES concerning my accreditation and procedures should Chris have a complaint concerning the work that we did.

Following this discussion I moved on to gathering more information via a semi-structured interview. In this phase of assessment my overarching aim is to gain knowledge concerning the psychological states that the athlete experiences before, during and after competitions and training. For me, the key psychological states that I look to find out more about are (in no order of importance) confidence, perceptions of control, anxiety levels, attention/concentration levels, motivation and commitment, enjoyment and positive emotions. This is because the literature on the impact of each of these on sporting performance and an athlete's experience is compelling enough to indicate that changes in these factors can produce changes in performance and enjoyment of sport (Lavallee, Kremer, Moran and Williams, 2004). My underlying philosophy (cognitive–behaviourist) is that these key mental states are, in turn, determined by (a) the habitual cognitive processes of the individual and (b) the environmental conditions which athletes encounter.

Given this philosophy, should it emerge in the interview that the performer has less than optimal psychological states (e.g. high anxiety, low confidence, concentration difficulties) I would seek to explore the cognitive and behavioural determinants of these states. In terms of the cognitive determinants of such states, I am interested in finding out about the client's view of the world, their theories of self (self-image, self-concept, self-identity) and their schemas (Hill, 2001) related to the world of golf (such as what it takes to be a successful golfer, what a good performance is). The aim is to uncover habitual, faulty or irrational thinking patterns (e.g. 'If I don't make my first putt of the day I am bound to have a poor day on the greens' or 'If I am not hitting the ball well I cannot score well'). In terms of the behavioural underpinnings of behaviour I am keen to find out the environmental triggers of maladaptive cognitions and emotions (e.g. the presence of other people, particular courses/holes) and also the extent to which adaptive behaviours, cognitions and emotions are created and/or reinforced by the environments in which the athlete performs and trains (Hill, 2001).

In order to elicit this information I attempt to ask very general questions concerning the client's experiences and practices within their sport. I have no pre-determined intake protocol (e.g. Taylor and Schneider, 1992) but I have found the writings of practitioners such as Davies and West (1991), Gardner and Moore (2005) and Vealey and Garner-Holman (1998) useful when considering information to seek and questions to ask to develop a deeper knowledge of the performer. My first line of questioning concerns the sporting history and current sporting involvement of the client. This includes when they started playing their sport, their current level of performance, their current training and pre-competition preparation practices. I will also ask my clients to describe their best and worst performances. In the case of Chris I was interested to discover more about his background in golf (how long he had been playing and his past achievements), his current involvement with golf (his handicap, the level of competitions he entered, his practice schedule, his contact with a coach) and any factors that may have influenced, were influencing or could influence his performance in the immediate future. These areas broadly resonate with what Taylor and Schneider (1992) would term athletic history and what Gardner and Moore (2005) would term contextual performance demands, skill level and situational demands information. I find starting with this rather general line of enquiry to have three main benefits. First, the young golfers that I have worked with find that these questions (which largely involve descriptions of events and the world of golf rather than their emotional and cognitive experiences in golf) are easy to answer and reduce some of the threat that can be inherent in their visit to a sport psychologist. Second, the general nature of the questions allows me to listen for clues regarding the schemas/beliefs the client holds and the role of the environment that I may be able to probe further with more questioning. Third, it is extremely useful information when it comes to matching my choice of intervention to the situation that the client finds him or herself in. For instance, in this case I found that Chris's school examinations were very important to him and would severely limit his time to either play golf or commit to mental skills training or psychology sessions.

As the interview progressed, it became evident that Chris was perhaps not as confident as he could be and, at crucial times (e.g. in the company of county officials or when attempting to 'close out' important club and county matches), he lacked self-belief. When issues like this arise in a consultancy my aim is to attempt to get as rich a picture of the athlete's experiences as possible. In these instances, I have found it useful to incorporate Davies and West's (1991) 'BASIC-ID' suggestions for developing further knowledge. Thus, when Christopher talked about times when he lacked confidence I was able to probe how this had affected his Behaviour, his Affect, his Sensations, his Imagery and his Cognitions (I rarely use the ID element of the guidelines, namely Interpersonal relationships and Diet). I was also able to include Chris's parents in the discussion at this point to get their opinions of Chris's behaviours and their perceptions of his responses to situations where he was low in confidence. This gave me a greater idea of how low confidence manifested itself in Chris and how it affected him on the golf course and on the practice ground.

The second method of information gathering I used with Chris was Bull, Albinson and Shambrook's (1996) Mental Skills Questionnaire (MSQ). The aims of this were (i) to assess Chris's current level of use of mental skills from which interventions could be built, (ii) to supplement the information obtained through the interview and (iii) to aid evaluation of any intervention. In other situations I have used the Test of Performance Strategies (TOPS: Thomas, Murphy and Hardy, 1999) to gather information from clients. However, given the age of the golfer and his relatively intermediate level of performance I decided that this might be overly long and complex for Chris. As I was also unsure of his level of intellectual ability at the time of planning the session, and did not want him to be overawed by such a lengthy assessment tool, I decided not to use this tool at this point in time. For me, although the MSQ may lack the psychometric properties of other assessment tools, it is short and easy to understand and I have often used it, as in this case, as a stimulus for further discussion about mental skills in general at the early stages of consultancy.

Analysis – aims of the intervention

Two key points emerged from the discussion between myself, Chris and his parents. The first issue was Chris's concern that he lacked self-confidence and belief in himself. This was confirmed by the results of his Mental Skills Questionnaire that revealed a low score (11/24) on the Self-confidence sub-scale and was also very evident during the interview. Through talking about his confidence the first thing that became apparent was that Chris's lack of confidence in his ability was exacerbated by a strong outcome orientation. From his responses in this session it seemed as though his predominant criterion for success was beating others. Thus although he reported having many good rounds and many good experiences playing golf, his confidence would be dented if he did not beat people he expected to beat or if he thought he was falling behind his peers (in fact, one of the catalysts for Chris seeking the services of a sport psychologist turned out to be that he had recently failed to be included in a regional training squad when other members of his club and county teams had been included). Research has established that adopting goals that focus on outcomes and defeating others, when coupled with low levels of perceived competence (self-confidence), can lead to maladaptive behaviours (e.g. low levels of effort) and cognitions (e.g. anxiety, cognitive interference, low levels of enjoyment; Roberts, 2001; Weiss and Ferrer-Caja, 2002; Roberts, Treasure and Conroy, 2007). Such an outcome orientation has also been associated with lower levels of self-confidence due to the athlete having a restricted chance of achieving successful outcomes (Roberts, 2001). Research has also shown that task orientation and involvement can be effectively enhanced in young athletes (Harwood and Swain, 2002). Thus, one of the fundamental aims of this consultancy for me was to try and increase Chris's task/process orientation and involvement in order to give him alternative ways of gauging success and so enhance his self-confidence, his enjoyment of golf and his intrinsic motivation.

Another determinant of Chris's lack of confidence was his excessively high expectations of himself. Chris displayed a strong perfectionist streak in the standards that he set for himself. What emerged from our conversations was that he expected himself to have statistics that even top tour professionals would be proud of. For instance, he felt that he should be holing 10 out of 10 putts from 3 feet and 6 feet and nine out of 10 putts from 10 feet and he reported feeling frustrated and low in confidence when he was not achieving these scores. However, inspection of the statistics of tour professionals (Pelz, 2000) indicates that tour professionals average nine out of 10 for

3 foot putts, seven out of 10 for 6 foot putts and only about five out of 10 for 10 foot putts. This indicated to me that his theory of what constituted good golf for a player of his age and handicap was not justified and that it was causing him to (a) become frustrated with himself, (b) be disappointed with performances that were objectively good and (c) retain low levels of self-confidence. Thus, it was clear that the intuitive theories of golf that Chris held were inaccurate and leading to potentially damaging psychological states.

Chris's lack of confidence also seemed to be creating greater anxiety before rounds (concentrating on how the opponents were playing and who was there watching, and worrying about these factors) and problems concentrating during it. This was supported by similar low scores on Anxiety and Worry Management (11/24) and Concentration Ability (9/24), and by the statements and observations of his parents. Chris also reported getting very nervous throughout the day of the competition and when on the first tee. Thus, the second focus of the intervention was giving Chris the mental skills to deal both proactively and reactively with anxiety and concentration problems.

Some of the contextual information was also very useful in formulating potential intervention decisions. First, Chris was clearly a very thoughtful and intelligent person. He was also doing a large number of school exams. The advantage of this was that I felt confident that if I asked Chris to do 'homework'[1] tasks for me he would diligently attempt to do them, completing the task and doing so in as full a manner as possible. The disadvantage was twofold. First, it meant that the time that Chris was going to be able to dedicate to any mental skills training that we decided upon may be affected. Second, it meant that at some point at the height of the golfing season Chris would have to decide on the balance of time he spent on golf practice and competition and time he spent on school work. As his school work was obviously important to him (and his parents were keen that he did well in his exams), this would mean that his expectations at this time would have to be managed carefully. In the past, I had observed golfers who could not accept that their performances were not improving (or were declining) when they had been forced to reduce the amount of practice they could do and the number of competitions they could enter.

The second element of contextual information that was important to me was the amount and nature of weekly practice that Chris engaged in. One of my perceptions as a sport psychologist working with golfers (and one shared by such sport psychologists as Richard Cox, 1997) is that practice is often not as effective as it could be in (a) fostering the skills needed to be successful in competition and (b) building the confidence of

---

[1]Normally I am loath to use such terms with junior athletes, especially when they are in the midst of school examinations. However, in this instance I felt it was something that Chris would be confident of doing well for me.

golfers. It is also often evident that many young golfers, when practising alone, often do not structure their practice in order to give themselves clear and unambiguous feedback. In Chris's case, it was my impression (from his descriptions of his practice habits and schedule, corroborated by his parents) that his practice sessions did not provide as much opportunity for reinforcement as they could have done or, perhaps, as much as Chris needed. For instance, when practising technical aspects of his game (e.g. working on swing changes) Chris would hit countless balls but would gather little feedback concerning whether he was executing the skill well or improving in his execution of his swing beyond that of the feel of the swing and the final location of the ball. Also, when practising specific elements of his game (e.g. chipping, mid-irons) Chris gave himself little opportunity to see how he was improving in these areas. Instead, when I asked how Chris knew that a training session had been productive he reported that it was purely down to how he felt he had been swinging the club. Thus, the aim of the consultancy was also to try, within the constraints of Chris's golf club, time and money, to enhance his practice environment to increase the opportunities for specific feedback and reinforcement.

6.3 Interventions and monitoring

The aim of the intervention was thus twofold. Firstly, it was to develop greater self-confidence within Chris. To do this the aim of the intervention was to (a) broaden Chris's view of what constituted success, (b) structure Chris's practice to ensure that he was experiencing success and positive reinforcement in practice and (c) modify Chris's expectations about his golfing performance so that they were more realistic and more conducive to raising his self-confidence. A secondary aim of the intervention was to give Chris the mental skills to improve his on-course concentration skills and to deal more effectively with on-course distractions and anxiety. In addition to this, as one of Chris's parents was present at every meeting (as is ethically and legally appropriate when dealing with a junior), they were given advice on how to reinforce the messages of each session.

Developing task/process orientation and building confidence

Chris was largely outcome-focused. However from the discussions we had, this was largely due to him being unaware of alternative ways of gauging his success rather than the attitude that winning was all that mattered. Chris and his parents appreciated the benefits of broadening his conception of what constituted a successful performance when we talked about the importance of process orientations (Roberts, 2001) and taking a long-term perspective of trying to identify areas for improvement. In addition to this we focused on two strategies to develop further a process orientation.

Performance profiling and goal-setting (for both competition and training)

Chris and I went through a performance profiling procedure (Butler and Hardy, 1992) whereby, as homework from our first session, Chris was asked to come up with as many areas as possible where a professional golfer needs to excel as possible. These were discussed at the second meeting and broadened as much as possible (therefore, through discussion, putting was broken down into 3, 6, 10 and 15+ feet putts). Following this, Chris rated himself on each of the resultant constructs from 1 to 10, with 10 being defined as the performance of a top England junior player (fortunately, there was such a golfer in Chris's county squad who served as a benchmark). The performance profile session was thus used as a means of discussing the process of performance and for finding areas of his game that Chris perceived himself to be weak on (by breaking each element of golf down as in the putting example, we also sought to make the weakness seem a smaller problem in the context of Chris's all-round game). Such sessions have been proposed by researchers (e.g. Butler and Hardy, 1992) to be a mechanism for enhancing the task orientation of athletes. On the basis of the profile we set goals for a number of areas of Chris's game. These goals were focused on the processes and performances underlying golf (e.g. improving sand saves, improving driving distance and accuracy, reducing the number of putts). This has been proposed to be an effective means of enhancing task involvement and through this enhancing self-confidence (Burton, 1989, 1992).

The goal-setting aspect of the session consisted of, firstly, talking to Chris about the impacts of improving certain aspects of his game (e.g. 'what would be the impact if you increased your accuracy off the tee?'). Following this we discussed how to set SMART goals (specific, measurable, adjustable, realistic and time-based) in order to achieve these improvements. However, for me the essential element of goal achievement is to develop clear and specific strategies for achieving those goals. Thus, Chris and I discussed goals for his training. First, we discussed planning Chris's practice sessions so that all elements of his game were practised over the period of a week (previously, Chris reported that he 'just turned up and hit some balls') and he recorded his practice session so that (a) he could monitor the practice he did and (b) he could see the amount of practice he did. Second, we discussed setting specific goals for practice that Chris could set for himself so that he covered all elements of his game and did a specified number of drills in targeted areas of his game. In order to facilitate these goals and to provide a greater level of unambiguous feedback on his progress over the winter months, Chris and myself worked on a number of skills tests as a basis of goal setting (see Figure 6.1 for an example). The skills tests were designed to monitor how effective Chris was on certain types of shots. The aim was for Chris to conduct the skills tests two or three times a week following his practice sessions and to record

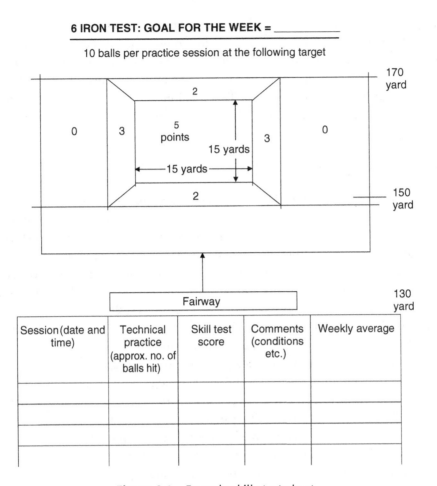

Figure 6.1 Example skills test sheet.

his score as a weekly average. Following the first week (baseline), Chris and I talked about setting specific goals for improving his performance in each of the skills tests and self-rewards for achieving them. Thus, the skills tests assisted in giving more structure to Chris's practice sessions and gave him the mechanism for receiving more specific feedback than his previous reliance on subjective impressions of how his training was progressing. The progress made on the skills tests was reviewed at every meeting to discuss issues concerning progress, lack of progress, ease of scoring and sensitivity of scoring system.

In addition to the goals we set for training sessions we also talked about setting goals for the rounds (both competitive and practice) that Chris played. This again focused on achieving performance (e.g. number of putts, number of fairways hit, number of sand saves) and process goals (e.g. keeping head down on putts, using pre-performance routines before every shot). Again, Burton (1989, 1992) provides support for the efficacy of this approach for confidence enhancement and further research (e.g. Kingston and Hardy, 1997; Filby, Maynard and Graydon, 1999) has indicated not only that the use of performance and process goals aid self-confidence, but also that they aid concentration. This fitted in well with the second aim of the intervention and, given Chris's intelligence and diligence, recording this information did not prove to be a problem.

Performance evaluation

We also used a derivative of Holder's (1997) Performance Evaluation Sheets. The exercise is relatively simple in that the performer is asked to provide a list of positive and negative aspects of each performance (negatives should never exceed positives). To aid the evaluation process we split the evaluation into categories (physical, psychological, technical, tactical and preparation) and Chris was also encouraged to reflect on the constructs from his performance profile when completing his evaluation form. In addition to this, Chris was also asked to record his three best shots of each round and to give as much detail about these as possible. The aim of this was to increase the amount of time Chris spent thinking about the good shots from his round.

The aims of encouraging Chris to adopt a more systematic approach to performance evaluation were multiple. Firstly it would provide, together with his profile, a further basis for goal setting (the aim here was to work on his weaknesses but also to highlight that he did have control over such weaknesses). Secondly, it would force Chris to take a more process-oriented approach by looking for aspects of his performances that were positive and in need of improvement rather than purely labelling a round as good if he beat others or played below his handicap and poor if he did not. Thirdly, it would provide a written record of Chris's positive experiences and so be something that he could draw confidence from before big events. It was also hoped that it would train him to draw out positives from even the worst performance. Fourth, it was hoped that this exercise would encourage him to evaluate and attribute his rounds to more controllable factors as this pattern of attribution has been proposed as being most likely to protect confidence levels (Bandura, 1997). This proposition has received research support in sport settings (Orbach, Singer and Price, 1999; Orbach, Singer and Murphy, 1997), although the findings of these studies must be treated with some caution as they have only examined attribution retraining in novice performers following failure. Finally, the information contained in the forms and the discussion of the evaluations provided further information for me to refine the intervention as it progressed.

Developing more realistic expectations

Throughout the intervention Chris and myself talked about his expectations and worked on ensuring they were based on objective information rather than incomplete and subjective information. In order to encourage Chris to think more about the beliefs that he held I used a variety of cognitive techniques. First, I provided direct challenges (Hill, 2001) to Chris's statements concerning his beliefs about golf. For instance, when Chris expressed the belief that if he did not make it to a certain handicap by the time of his sixteenth birthday he would never make it as a top player I was able to challenge this belief and provide a number of examples of successful golfers who had taken the game up late or who were 'late developers'. To this end, my experience of working within golf was greatly beneficial to the power of my message to Chris. I also attempted to use collaborative empiricism (Hill, 2001) to challenge Chris's self-statements and theories of golf. This involved setting Chris homework to find evidence concerning his beliefs. For instance, when discussing his beliefs about the performance of top golfers (which informed Chris's expectations of his own play), I asked Chris to complete a 'quick quiz' on what statistics he felt he should be getting and what statistics they felt the tour professionals would get. The answers provided by Chris revealed that his expectations were excessive. Thus, I gave Chris homework to find statistics out and to set himself challenging goals based on this new, more relevant, level of expectation. In a similar vein, I also asked Chris to examine his on-course statistics to challenge faulty beliefs such as 'if I start poorly I never play well', 'I always mess that hole up' and 'I never play the final holes well in important matches'.

Dealing with competitive anxiety and concentration difficulties

The second issue discussed was dealing with competitive anxiety and distractors on the course. Chris indicated that he often played shots with negative thoughts in his mind or with task-irrelevant thoughts in his mind. Through discussion we agreed to examine day-of-competition routines (in order to increase perceptions of control leading up to the competition) and pre-shot routines and also to look at ways of boosting confidence on the day of competition. The aim of all that we talked about was to give Chris positive and relevant thoughts to be thinking in an attempt to block out negatives. It was also the aim to enhance Chris's coping or ameliorative (Bandura, 1997; Feltz, Short and Sullivan, 2008) self-efficacy (his confidence in his ability to cope at key moments in his game such as on the first tee, when selectors were watching and when closing out matches).

Pre-round routine

The pre-round routine was aimed at giving Chris a sense of control over his day and ensuring that everything that had to be done to increase the likelihood of a successful performance was done. This is an approach advocated in numerous sport psychology texts (Bull *et al.*, 1996; Wilson, Peper and Schmid, 2006). Although the evidence supporting the efficacy of pre-competition routines is largely experiential (e.g. Bull *et al.*, 1996), the aim of the pre-round routine was to ensure that Chris (a) had sufficient time to warm-up both physically and mentally, (b) avoided distractions and anxiety-inducing stimuli (in the form of fellow competitors and clubhouse advice givers), (c) perceived himself to be in control of the pre-round period of time and (d) focused on doing things that he felt facilitated good performances. The pre-round routine was based on Chris's reflections on his preparation before good rounds in the past, on his reflections of the stimuli that distracted him or worried him before rounds and my suggestions for managing his environment based on some of the strategies that previous clients had used. From this discussion, I asked Chris (with his parents monitoring him) to implement the pre-round routine prior to his next practice round. This served as a chance to practice his routine prior to using it in a competition and allowed us to discuss any changes that may have been needed.

Pre-shot routine

The pre-shot routine was designed to get Chris to focus fully on each shot, to be confident in his choice of shot and so fully commit to each shot. The efficacy of the use of pre-performance routines in golf has received support in the research literature (for reviews see Cohn, 1990; Lidor, 2007) and it was evident from my discussions with Chris that he believed (through his knowledge of the psychology of golf) that it was an area that he could improve on in his game. The aim of the pre-shot routine was to integrate a number of psychological skills, thought processes and behaviours that Chris believed led to more consistent ball-striking and shot making. The pre-shot routine that was developed initially, although subsequently refined and periodically updated, was as follows:

1. At the ball – MAKE DECISION (consider distance, wind, lie, obstacles, strategy) – DEEP BREATHS (breath in to a count of 4, hold for 2, out slowly to a count off 4) – 'JUST ANOTHER SHOT'.

2. First practice swing – think about lie and distance.

3. Second practice swing.

4. Stand behind ball – PICK TARGET – check grip.

5. Think about finishing point – VISUALIZE THE BALL LANDING WHERE YOU WANT IT TO.

6. Check stance (shoulder and feet) – 'IT WILL GO WELL'.

7. JUST HIT IT.

Following agreement on this as a suitable pre-shot routine, Chris and I agreed some goals for the progressive use of it in practice and then in competition. Thus, Chris moved from spending time practising it without hitting balls, to using it in 'free practice', to using it when performing his skills tests, to using it in practice rounds until he finally began to integrate it into his competitive rounds. By this stage, Chris felt comfortable with the pre-shot routine and was happy with the timing and content of the various stages.

6.4 Evaluation of interventions

Following Anderson, Miles, Mahoney and Robinson's (2002) suggestions, four effectiveness indicators were considered when evaluating the consultancy with Chris. These were psychological skills, athlete's responses to support, performance and quality of support. From my perspective, and because of the generally short lifetime of the consultancy to this stage, I was more concerned with the development of psychological skills, adaptive cognitive processes and attitudes towards sport psychology than performance (although this became more relevant when the consultancy was evaluated again at the end of the golf season). In order to gather information about these indicators, a number of strategies were used. These were informal discussions with Chris, informal (and separate) discussions with Chris's parents, conversations with the county organizer who had initially recommended my services to the family, asking Chris to complete Partington and Orlick's (1987) Consultant Effectiveness Questionnaire (CEF) and by asking Chris to complete a second MSQ.

The informal discussion with Chris was, in general, positive. Chris reported being pleased with the overall programme, particularly with the evaluation sheets and the pre-shot routine. Chris indicated that he felt much more confident and in control of his game most of the time. What struck me was that Chris's attitude to sport psychology was very good and he was keen to continue to use and refine the skills and strategies that we had talked about in the consultancy to that point. He was also keen to develop the mental side of his game further. Over the course of the intervention Chris also showed an improvement in his handicap (from 4 to 2) and performed well in a number of junior open competitions in his county, winning one. However, what pleased Chris most was his increased awareness of the impact of his self-beliefs and his theories of golf. By this stage of the consultancy it was pleasing to see Chris check himself mid-sentence to question the utility of certain statements that he made which had reflected some of his cognitive processes. This provided me with some evidence that the cognitive techniques that I had employed with Chris were having an effect on him. Chris also re-rated himself on the mental skills questionnaires, which mirrored his comments regarding his improvements. Specifically this revealed an improvement (of between two and four points) in the subscales of Mental Preparation, Self-Confidence, Anxiety and Worry Management, Concentration Ability and Relaxation Ability.

Despite the improvements in his game and his mental approach, Chris still had concerns about his ability to remain calm in the face of bad shots on the course and reported being intimated by opponents at certain events. This has since been covered in other sessions but, in hindsight, could have been covered in this intervention programme. Chris's responses to the CEF also revealed a very positive opinion of the sessions that we conducted with all responses indicating his satisfaction with the work that we did together. However, Chris did note that I could have been more involved in the development of the skills outside of the consultancy sessions. This resonated strongly with my reflections concerning how I could have been more effective.

Reports from Chris's parents and the county organizer were also encouraging. They reported that Chris seemed much more focused on his golf, more confident in his abilities and more balanced in his responses to successes and failures. The county organizer, who had not been involved in the consultancy, reported that he had noted that Chris had developed a consistent pre-shot routine and that he was more confident in himself than he had been previously. Chris's parents did, however, note that Chris still occasionally reverted back to his old self at times of greatest pressure. Although this was not that surprising given the lifespan of the consultancy at this stage, it was something that I was keen to continue to work on with Chris.

6.5 Evaluation of consultant effectiveness/reflective practice

When reflecting on my own effectiveness and the quality of the service that I provide, I have found Gibbs's (1988) staged model of reflection a useful guide. This has previously been suggested as a useful framework for sport psychologists engaging in reflections on their practice (Anderson, Knowles and Gilbourne, 2004). However, rather than extensive journal keeping, concept mapping or ethnodrama (see Anderson *et al.*, 2002 for further suggestions), my reflections have tended to occur more informally as I write summary letters to clients (a practice that was continued with Chris) and in informal discussions with colleagues and peers concerning the work that we are engaged in.

My overall reflection on the consultancy process with Chris was positive. Chris's attitude, intelligence and willingness to reflect on his own performances made the sessions with him enjoyable from my perspective and ensured that I was confident in dealing with him and his parents. The fact that (a) I had been recommended to the family by someone they trusted and who had a long history of involvement with the English Golf Union and (b) that I myself was working with the EGU and had a good understanding of the world of junior golf meant that I had little difficulty in establishing credibility and rapport with Chris and his parents. This was enhanced with a very professional initial session where I outlined clearly the roles that myself, Chris and his parents would fulfil in the consultancy process. In addition, the intervention consisted of a broad range of practical activities that Chris was able to comprehend quickly and, with the help of his parents, integrate into his golfing practices effectively.

However, the reflection process did highlight a number of learning points for me. The main learning points were (a) the importance of spending more time working with/observing the athlete in his/her training and competition environment, (b) the potential impact of cognitive techniques in my practice and (c) the wasted opportunity to involve the parents more in the consultation.

The importance of spending more time working with/observing the athlete in their training and competition environments

I feel I should have made greater attempts to see Chris in competition. Attending practices and competitions has been proposed as a very good strategy for enhancing rapport and client–practitioner relationships (Ravizza, 1988) as well as the obvious benefits of providing a further opportunity to assess the client and move away from reliance on self-report information (Taylor, 1995). Unfortunately, due to my work commitments (during the week with my university and during the weekend with other consultancy work) and Chris's school work, it was difficult to arrange a time to visit Chris at his golf club. However, I feel as though I could have made more of an effort to attend a training session. This would also have afforded me more of an opportunity to discuss issues such as Chris's use of his pre-shot routine and to get a better idea of the facilities at his disposal for practice. Having since observed Chris playing golf I feel I would have been better placed to comment on his performances and to suggest more individually tailored interventions. This has been something that I have tried to achieve early in more recent consultancies and I feel that it has not only given me a greater insight into the athletes I work with, but has also enhanced the working relationship between myself and clients by demonstrating my commitment to them.

The potential impact of cognitive techniques in my practice

During the consultancy with Chris (which occurred at the same time as I was implementing more cognitive techniques with other clients), I was impressed with the effects of the cognitive techniques I was using. The activities (e.g. direct challenges, collaborative empiricism) that I attempted with Chris seemed to have a very positive impact on his belief systems (this was also observed when I finally saw Chris in competition). Although other elements of the intervention (or even factors outside of the intervention such as maturation) could have been responsible for such changes in attitudes, the activities struck me as a useful addition to my consultancies and highlighted a continuing professional development need/opportunity for myself. I have since tried to develop my knowledge of cognitive approaches to performance enhancement.

The wasted opportunity of involving Chris's parents to a greater extent

I was pleased with the way that I managed the sessions with the presence of Chris's parents. I feel that establishing ground rules about their input from the start helped to create a comfortable environment for Chris as he knew that he was in control of the sessions and his parents could not dominate. It helped enormously that Chris and his parents were intelligent people who could quickly grasp the concepts that we were discussing. Despite this, I feel that I could have 'used' the parents to a greater extent to reinforce the messages that were covered in the sessions. Harwood and Swain (2001, 2002) have shown the benefits of educating and incorporating parents into sport psychology support services. Although the parents were present and responded well, I do feel that I could have given them greater help in reinforcing the sessions. At one stage, I did consider the idea of offering to have a separate session with Chris's parents to discuss issues that emerged from the consultancy (such as how they could help Chris reflect on his performances and the most appropriate timing of such assistance), but did not go through with this as I was concerned with how Chris would react. However, more careful consideration of how to give the parents more guidance on how to assist Chris could have reinforced the messages I was attempting to give to Chris.

6.6 Summary

Self-confidence is commonly accepted as being associated with sporting performance and the quality of an athlete's sporting experience (Feltz et al., 2008). The preceding case study sought to outline one approach to enhancing self-confidence by targeting the specific antecedents of low levels of self-confidence (high ego-orientation, impoverished opportunities for self-reinforcement of performances) and devising an intervention around these. The approach that was settled on combined cognitive and behavioural approaches which aimed at, firstly, examining the core beliefs and schemas held by the athlete and trying to produce more adaptive cognitions and, secondly, examining ways in which the athlete's environment (in this case practice) could be manipulated to provide more opportunities for success to be achieved, recognized and attributed to the performer. Furthermore, strategies (i.e. pre-performance and pre-shot routines) were also put in to place to give the performer some control over the consequences of lower levels of self-confidence (anxiety, attention control). Although the interventions could have been more effective had I spent more time with the client on the practice ground and on the course and/or utilized the parents to a greater extent, subjective and objective assessment showed that the consultancy with Chris was effective.

Questions for students

1 What factors may enhance the nature of the client–practitioner relationship?

2 What practical activities should/could sport psychologists integrate into their working methods to enhance the client–practitioner relationship?

3 Identify antecedents or determinants of self-confidence in sport?

4 Discuss strategies for enhancing self-confidence in sport, matching these strategies to specific antecedents of self-confidence.

5 Explain how expectations may influence psychological functioning. How can sport psychologists ensure that performers develop realistic expectations?

References

Anderson, A.G., Knowles, Z. and Gilbourne, D. (2004) Reflective practice for sport psychologists: concepts, models, practical implications and thoughts on dissemination. *The Sport Psychologist* **18**, 188–203.

Anderson, A.G., Miles, A., Mahoney, C. and Robinson, P. (2002) Evaluating the effectiveness of applied sport psychology practice: making the case for a case study approach. *The Sport Psychologist* **16**, 433–454.

Bandura, A. (1997) *Self-Efficacy: The Exercise of Control*. New York, Freeman.

Bull, S.J., Albinson, J.G. and Shambrook, C.J. (1996) *The Mental Game Plan: Getting Psyched Sport*. Sports Dynamics, Eastbourne.

Burton, D. (1989) Winning isn't everything: examining the impact of performance goals on collegiate swimmers' cognitions and performance. *The Sport Psychologist* 3, 105–132.

Burton, D. (1992) The Jeckyll/Hyde nature of goals: reconceptualising goal setting in sport. In: Horn, T. (ed.), *Advances in Sport Psychology*, pp. 267–297. Human Kinetics, Champaign, IL.

Butler, R.J. and Hardy, L. (1992) The performance profile: theory and application. *The Sport Psychologist* 6, 253–264.

Cohn, P.J. (1990) Preperformance routines in sport: theoretical support and practical implications. *The Sport Psychologist* 4, 33–47.

Cox, R. (1997) The individual consultation: the fall and rise of a professional golfer. In: Butler, R.J. (ed.), *Sport Psychology in Performance*, pp. 129–146. Butterworth, Oxford.

Davies, S. and West, J.D. (1991) A theoretical paradigm for performance enhancement: the multimodal approach. *The Sport Psychologist* 5, 167–174.

Feltz, D.L., Short, S.E. and Sullivan, P.J. (2008) *Self-Efficacy in Sport*. Human Kinetics, Champaign, IL.

Filby, W.C.D., Maynard, I.W. and Graydon, J.K. (1999) The effect of multiple-goal strategies on performance outcomes in training and competition. *Journal of Applied Sport Psychology* 11, 230–246.

Gardner, F.L. and Moore, Z.E. (2005) Using a case formulation approach in sport psychology consulting. *The Sport Psychologist* 19, 430–445.

Gibbs, G. (1988) *Learning by Doing: a Guide to Teaching and Learning Methods*. Oxford Brookes University, Further Education Unit, Oxford.

Harwood, C. and Swain, A. (2001) The development and activation of achievement goals in tennis: I. Understanding the underlying factors. *The Sport Psychologist* 15, 319–341.

Harwood, C. and Swain, A. (2002) The development and activation of achievement goals within tennis: II. A player, parent and coach intervention. *The Sport Psychologist* 16, 111–137.

Hill, K. (2001) *Frameworks for Sport Psychologists: Enhancing Sport Performance*. Human Kinetics, Champaign, IL.

Holder, T., (1997) A theoretical perspective of performance evaluation with a practical application. In: Butler, R.J. (ed.), *Sports Psychology in Performance*, pp. 68–86. Butterworth-Heinemann, Oxford, UK.

Kingston, K.M. and Hardy, L. (1997) Effects of different types of goals on processes that support performance. *The Sport Psychologist* 11, 277–293.

Lavallee, D., Kremer, J., Moran, A.P. et al. (2004) *Sport Psychology: Contemporary Themes*. Palgrave-Macmillan, Basingstoke.

Lidor, R. (2007) Preparatory routines in self-paced events: do they benefit skilled athletes? Can they help the beginners? In: Tenenbaum, G. and Eklund, R.C. (eds), *Handbook of Sport Psychology* (3rd edn), pp. 445–465. John Wiley & Sons, Hoboken, NJ.

Orbach, I., Singer, R.N. and Murphy, M. (1997) Changing attributions with an attribution training technique related to basketball dribbling. *The Sport Psychologist* 11, 294–304.

Orbach, I., Singer, R.N. and Price, S. (1999) An attribution training programme and achievement in sport. *The Sport Psychologist* 13, 69–82.

Partington, J. and Orlick, T. (1987) The sport psychology consultant evaluation form. *The Sport Psychologist* **1**, 309–317.

Pelz, D. (2000) *Dave Pelz's Putting Bible: The Complete Guide to Mastering the Green*. Doubleday Publishing, New York.

Pocwardowski, A., Sherman, C.P. and Henschen, K.P. (1998) A sport psychology service delivery heuristic: building on theory and practice. *The Sport Psychologist* **12**, 191–207.

Ravizza, K. (1988) Gaining entry with athletic personnel for season-long consulting. *The Sport Psychologist* **2**, 243–254.

Roberts, G.C. (2001) Understanding the dynamics of motivation in physical activity: the influence of achievement goals on motivational processes. In: Roberts, G.C. (ed.), *Advances in Motivation in Sport and Exercise* (2nd edn). pp. 1–50. Human Kinetics, Champaign, IL.

Roberts, G.C., Treasure, D.C. and Conroy, D.E. (2007) Understanding the dynamics of motivation in sport and physical activity: an achievement goal interpretation. In: Tenenbaum, G. and Eklund, R.C. (eds), *Handbook of Sport Psychology* (3rd edn). pp. 3–30. John Wiley & Sons, Hoboken, NJ.

Salacuse, J. (1994) *The Art of Advice*. Times Books, New York.

Taylor, J. (1995) A conceptual model for integrating athletes' needs and sport demands in the development of competitive mental preparation strategies. *The Sport Psychologist* **9**, 339–357.

Taylor, J. and Schneider, B.A. (1992) The sport-clinical intake protocol: a comprehensive interviewing instrument for applied sport psychology. *Professional Psychology: Research and Practice* **4**, 318–325.

Thomas, P., Murphy, S. and Hardy, L. (1999) The test of performance strategies: development and preliminary validation of a comprehensive measure of athlete psychological skills. *Journal of Sports Sciences* **17**, 697–711.

Vealey, R. and Garner-Holman, M.G. (1998) In search of psychological skills. In: Duda, J.L. (ed.), *Advances in Sport and Exercise Psychology Measurement*, pp. 433–446. Fitness Information Technology, Morgantown, WV.

Weiss, M.R. and Ferrer-Caja, E. (2002) Motivational orientations and sport behaviour. In: Horn, T. (ed.), *Advances in Sport Psychology*, pp. 101–184. Human Kinetics, Champaign, IL.

Wilson, V.E., Peper, E. and Schmid, A. (2006) Strategies for training concentration. In: Williams, J.M. (ed.), *Applied Sport Psychology: Personal Growth to Peak Performance* (5th edn). pp. 404–422. McGraw-Hill, New York.

SECTION B

Working with Teams and Squads

7

Delivering Sport Psychology to Olympic Rowers: Beyond Psychological Skills Training

Chris Shambrook

K2 Performance Systems Ltd, Reading, UK

7.1 Introduction/background information

The purpose of this chapter is to provide the reader with an insight into some specific elements of consultancy that have proved to be particularly useful for enhancing the performance mindset of a group of rowers generally and also specific crews. The longitudinal nature of the consultancy role provides some useful concepts related to the philosophy of delivering sport psychology, which will be highlighted within the chapter. 1 have been the sport psychologist for the Great Britain Rowing team since early 1997. Much has been learned about the sport and the delivery of sport psychology within an elite setting during this time. The focus of support is primarily geared towards the Olympic Games, but there are also annual World Championships and International regattas on the performance agenda. During the course of this chapter contextual and philosophical information will be detailed in order that the applied work reported can be appraised for its potential to be utilized in other contexts, sports or situations. It should also be noted that this chapter focuses on work that is 'beyond psychological skills training' to highlight the application of psychology generally to

Applied Sport Psychology   Edited by Brian Hemmings and Tim Holder
© 2009 John Wiley & Sons, Ltd

a sporting context. However, the theoretical and applied development required to deliver psychological skills training in a practical and impactful manner is an essential foundation upon which the current chapter is based. Without the grounding in core sport psychology techniques, it would not have been possible to introduce the more general psychological principles discussed here.

Contextualizing the consultancy

There is a contemporary and historical success associated with the Great Britain Rowing Team. Going back to 1984, and Sir Steve Redgrave's first Olympic Gold Medal on Lake Casitas, through to the success of two gold, two silver and two bronze medals at the Beijing Olympics, it is clear that expectations of delivering at the Olympics are always high for GB Rowing. This track record of success and an expectation for ongoing success brings a particular set of requirements when considering any element of sport science and medicine support being delivered for the coaches and athletes. Therefore, this backdrop has had a very strong influence upon my mindset and approach as a practitioner. Had I been engaged in a sport that was attempting to begin a winning record at the Olympics, the general philosophy behind my role and approaches taken may have been different.

My preference as a practitioner is to work from a cognitive–behavioural perspective. Within the general demands of high-level sport, this approach allows me to consider the day-to-day thoughts and actions required to develop and gain control over delivering performance. Additionally, the cognitive–behavioural approach allows me to effectively conceptualize the specific demands of the performance environment itself, and thus ensure that athletes have the requisite self-awareness and self-management skills to fulfil potential when it matters most.

The work that is reported here represents some of the choices made regarding how to use my allotted annual commitment of 60 days. Having been immersed within the culture of the sport for some time now, this presents totally different consultancy opportunities and requirements when compared with a one-off intervention with a particular team or individual. Equally, the work is able to take place as a result of many previous interventions.

The 60 days of consultancy within a 12-month period result in sporadic face-to-face consultancy. Coaches and athletes have access to the sport psychologist at a mixture of domestic training days, one or two international training camps, one international regatta and the annual World Championships or Olympic Games. Athletes and coaches also have telephone and e-mail access throughout the year purely as a function of having a sport psychologist as part of the entire support team. For a full World Championship team there would typically be up to 12 coaches and 60 athletes at the event. These numbers serve as a useful guide for the typical numbers under consideration within the role of sport psychologist throughout a season. Therefore, everything you read here has been delivered within these constraints and is based upon relationships built up over time.

From a context perspective, it is also important to note that the coaches I have worked with are all highly effective 'psychologists' in their own right, who manage the day-to-day psychology challenges with a similar level of expertise with which they design and deliver training programmes, coach technique and provide tactical direction. The coaches are very capable of delivering goal-setting, visualization, concentration training and confidence building to a very good level and as such simply being able to deliver psychological skills training as the sport psychologist would add little to the performance environment. Therefore, as a sport psychologist, my role has to 'add value' to this existing expertise base and the fact that I have a Ph.D. in psychology and 'expert' training in sport psychology does not mean that I have the monopoly on things psychological within the rowing team. My qualifications and experience do mean that I am expected to deliver my work in a way that makes a difference to the results that are delivered in competition. Just as the coaches and athletes are judged by their results, so too is the whole support team. As someone who passionately believes in the differences that effective psychology can make, I have no problem being expected to deliver results. Any different reaction would mean that I believe sport psychology is a 'nice to have' service that may or may not make a difference – hardly a glowing tribute for employing my professional services.

Finally, in terms of positioning the following work, it is important to note that I have made significant mistakes during my work within rowing, which are topics for a different chapter. Therefore, the concepts being presented here are ones that have been particularly interesting and effective, but I do not want to misrepresent my performance simply by leading the reader to believe that all my work is of this quality. The mistakes that have been made have been influential in all the good work that has resulted, but the journey has by no means been a faultless one.

7.2 Initial needs assessment

In the present context, the notion of needs assessment is an interesting one. With ongoing work behind me there is not a specific needs analysis process that I carried out with the athletes and coaches in order to determine a specific set of interventions, as might be expected from linear models of sport psychology consultancy (e.g. Thomas, 1990). A needs analysis of a formal nature might most typically be carried out for a one-off project or when a psychologist is called in to solve a problem. Immersion into the sport for an extended period of time results in a constant process of responding to specific needs, but more importantly, shaping a view of activities to carry out that will proactively allow athletes and coaches to systematically improve their psychological performance. The complex and changing nature of ongoing consultancy was expertly articulated by Hardy, Jones and Gould (1996), who demonstrated the need to have a dynamic approach to deliver expert consultancy within a 'collaborative and non-sequential' (p. 290) context.

In the nonlinear consulting reality that I deliver within, I have key questions that I use to help assess appropriate initiatives to put in place at any point in time. These are:

- Have I got to the point where I can be 100% sure that there will never be a psychological explanation for underperformance in training or competition for all individuals and crews?

- What is the next step I need to take to ensure athletes and coaches are systematically becoming the most psychologically prepared performers in the world?

- What do I need to do to ensure that self-awareness and self-management are constantly being improved?

It is clear from these questions that the specific bias of the needs analysis is primarily to move all high-performing individuals towards an optimum position with regards to their performance psychology. The questions result in a proactive mindset for myself as a practitioner and ensure that I am using my expert judgement to make decisions about the appropriate course of action to take at any point within a season or Olympiad.

In addition to the key questions, I can clearly identify key sources of information that are utilized to form expert judgements on the appropriate targeting of my time and focus of my work. The following are critical components of the needs analysis:

- Annual reporting to evaluate the impact of delivered work in combination with identification of actions to focus on in the coming year.

- Discussions with chief coaches about areas for further improvement within squad psychology.

- Discussions with athletes about performance psychology for training and competition.

- Observations of athletes and crews in training.

- Observations of athletes and crews in competition.

- Discussions with other sport science and sports medicine staff regarding areas of psychology relevant to their areas of expertise and influence.

- Information gleaned from one-to-one and crew consultancy sessions.

- Discussions with other sport psychology practitioners about their activities and approaches within different sports.

- Discussions with practitioners working outside of sport about training programmes being delivered for clients in areas of psychology.

As a result of these approaches, I would actually see my consultancy work within GB Rowing as a longitudinal action-learning project. 'Action learning is a real-time learning experience that is carried out with two equally important purposes in mind: meeting an organizational need and developing individuals and groups' (Rothwell, 1999, p. 5). Therefore, there is ongoing appraisal of opportunities to take action, followed by specific actions to respond to those opportunities, followed by appraisal and assessment of the result of those actions with a view to finding out what has been learned and what the most appropriate next steps are.

7.3 Interventions and monitoring

There are two specific activities reported within this chapter that represent the result of decisions made from the information gathered. One is a global approach that attempted to create a mindset shift within a whole rowing squad. The other is a specific tool that was developed to help crews prepare for competition. These varied activities have been deliberately chosen to demonstrate the breadth of the nature of work that has been carried out.

The first activity focused upon developing self-awareness, self-management, aware-ness of others, and management of others through the application of the Myers Briggs Type Indicator (MBTI) theoretical framework (Myers, 1962). This choice of activity resulted from the observation that no time had been spent explicitly addressing the psychology of working in groups and maximizing the presence of inevitable person-ality differences. Athletes and coaches were acutely aware of the day-to-day impact of personality differences on the effectiveness of working relationships, but no systematic steps had been taken to provide some practical advice on how to optimize harmony. Not only could this approach be of benefit from a squad point of view, but once intro-duced, the MBTI framework could also be used to help individual crews maximize the personality combinations in both preparation and racing. This work was initiated in the February of year three within an Olympic cycle, which is approximately 6 months before the World Championships.

The second activity focuses upon the adaptation of the Johari Window (Luft and Ingham, 1955) to develop a shared view of the specific performance attributes of the crew. This particular approach was decided upon as a result of identifying a need to create a simple way of capturing the 'performance identity' of the crew in readiness for competition, by using all of the shared knowledge of a crew, from within a crew. Creating coherent thinking amongst teams is an important area of consideration in many team contexts (Klimoski and Mohammed, 1994) and has particular importance with rowing as opportunities for communication are limited within performance. Therefore, finding ways to created shared views on performance factors and processes is worthwhile. At specific points within the competition calendar crews have differing amounts of competitor- and self-intelligence ahead of racing (i.e. what we know about the competitors and what we know about ourselves). Finding a way of creating a shared view of the competitive attributes of the crew in advance of these races can help to maximize competitive edge. In the particular case reported here, an adaptation of the Johari Window was decided upon to assess the extent to which this assisted the development of shared thinking for collective performance execution. The timing of this intervention was focused around racing and refers to utilization between May and August of the competitive season.

Myers–Briggs type indicator and raising awareness

Having identified that there would be benefit in carrying out a session to promote greater harmony within the squad in question through increasing understanding of personality dynamics and interactions, a recommendation was made to the chief coach to carry out a highly practical session utilizing the framework provided by the MBTI. Fortunately, the chief coach was open to this kind of approach and he has a great track record of utilizing sport psychology input alongside all of the other support services, in innovative ways, with a view to getting a performance improvement. Receptivity for such a session was an essential part of the ultimate success and, in all cases of sport psychology provision, the backing of the coaching staff is critical in the ultimate acceptance and utilization of work. With the agreement of the chief coach, a session was scheduled to begin the personality-based awareness raising. As a qualified MBTI practitioner, I was able to facilitate the session myself using all of the appropriate support materials.

The explicit purpose of the session was to introduce the squad to the concept of personality preferences to help them understand interactional dynamics within crews, between athletes and coaches, and within the squad generally. The MBTI session was about understanding individual differences and similarities present in the squad and exploring how these created specific challenges and opportunities within this specific high-performance environment. In terms of positioning the session with the coaches and athletes, there were two particular benefits that helped with the ultimate success of the work. First, the session was not about psychological skills training, it was about human interaction, regardless of context, and so was of interest to the athletes and coaches as people. Second, the session was carried out practically in an interactive manner, allowing the relevance of the concept to be identified for the specific context the rowers existed within. Rather than being recipients of a 'seminar'-style education session, the athletes and coaches were active participants in the whole session.

Background detail to MBTI

Before describing the session in detail, some background information about the MBTI is warranted. Comprehensive descriptions of the history and development of the MBTI from Carl Jung's typological theories of personality are available (Briggs-Myers, McCauley, Quenk and Hammer, 1998). Primarily, when considering raising awareness about personality dynamics, the MBTI introduces people to four distinct pairs of personality preferences, which can be briefly described as: how you prefer to interact with the world around you (Introversion or Extroversion); what kind of information you prefer to attend to (Sensing or Intuition); your preferred decision-making approach (Thinking or Feeling); and your preferred way of managing the world around you (Judging or Perceiving). A brief description of the pairings is provided in the form that was used with the group in question (see Figure 7.1). The four pairings are typically referred to in terms of the following letter pairings: I–E, S–N, T–F and J–P. Within the theoretical framework of the MBTI, the four pairings are considered to exist as dichotomies, with each individual having a preference for which of the two dichotomies within each pairing they most readily associate with. The MBTI simply reflects the degree to which someone recognizes how clear they are of their preference, rather than identifying whether someone is high or low in given type. Therefore, after assessment (either via questionnaire or through guided exercises), people end up with a four-letter combination capturing each of their preferences. There are 16 different combinations of the four letters, each of which helps outline preferred personalities as a function of the interaction between the different elements present. These are typically presented in a 'type table', as can be seen in Figure 7.2

Some key philosophical points are connected with the MBTI that are always communicated to those individuals being introduced to it. These are:

- The MBTI indicates preferred personality type and not the strength of a personality trait.

- There are no good or bad, right or wrong types; only differences, when considering the pairings and final type.

Extraversion–Introversion Dimension

Where do you prefer to focus your attention? Where do you get your energy?

Extraversion	Intraversion
Focused on the world outside you	Focused on their inner world
Prefer to communicate by talking	Prefer to communicate in writing
Work out ideas by talking them through	Work out ideas by reflecting on them
Learn best through doing or discussing	Learn best by reflection and 'mental' practice
Have broad interests	Focus in depth on their interests
Sociable and expressive	Private and contained
Readily take initiative in work and relationships	Take initiative when the situation or issue is important to them

very sure   quite sure   not sure   quite sure   very sure

Sensing–iNtuition Dimension

How do you prefer to take in information?

Sensing	iNtuition
Oriented to the present realities	Oriented towards the future
Factual and concrete	Imaginative and verbally creative
Focus on what is real and actual	Focus on patterns and meaning in data
Observe and remember details	Remember details when they relate to a pattern
Build carefully and thoroughly towards conclusions	Move quickly to conclusions, follow hunches
Understand ideas and theories through 'doing'	Want to clarify ideas and theories before putting them into action
Trust experience	Trust inspiration

very sure   quite sure   not sure   quite sure   very sure

Thinking–Feeling Dimension

How do you make decisions?

Thinking	Feeling
Analytical	Empathetic
Use cause and effect reasoning	Guided by personal values
Solve problems with logic	Assess impacts of decisions on people
Strive for an objective standard of truth	Strive for harmony and positive interactions
Reasonable	Compassionate
Can be 'tough minded'	May appear tender hearted
Fair – want everyone treated equally	Fair – want everyone treated as an individual

very sure   quite sure   not sure   quite sure   very sure

Judging–Perceiving Dimension

How do you deal with the world around you?

Judging	Perceiving
Scheduled	Spontaneous
Organise their lives	Flexible
Systematic	Casual
Methodical	Open-ended
Make short-and-long term plans	Adapt change course
Like to have things decided	Like things loose and open to change
Try to avoid last minute stress	Feel energised by last minute pressure

very sure   quite sure   not sure   quite sure   very sure

Figure 7.1   Descriptions of the four Myers–Briggs type indicator personality elements.

ISTJ	ISFJ	INFJ	INTJ
ISTP	ISFP	INFP	INTP
ESTP	ESFP	ENFP	ENTP
ESTJ	ESFJ	ENFJ	ENTJ

Figure 7.2 The Myers–Briggs type indicator type table.

- Individuals are their own best judges of type; hence a 'best-fit' process is carried out to ensure that each individual agrees with and owns their type. They are not simply informed of their type through imposed questionnaire results.

- MBTI results should never be used for selection purposes and should be entered into voluntarily.

Specifics of the practical MBTI session

For the purposes of raising self-awareness in the rowing squad, I decided that I would use a practical set of exercises to introduce and explore the MBTI framework rather than implement a full individual analysis utilizing questionnaires and exercises. The primary, applied aim of this session was to increase awareness of personality dynamics through the whole squad. The squad in question numbered 12 athletes and three coaches (two of whom were present for the session). Therefore, a live activity involving everyone, athletes and coaches, would achieve this goal far more effectively than having everyone complete their own questionnaire in isolation of their team-mates and coaches. For me this was a great example of keeping applied impact as my 'prime driver', when there might be other criteria I could choose to focus on that would actually hinder the quality of the message delivered to the coaches and athletes.

This session was a rare opportunity to have the entire squad in question for some 3 hours. Given the importance of high-volume, on-water training, my access to the athletes is typically grabbed between training sessions on the water, or towards the end of the training day, when more often that not the athletes would rather be getting home and resting. Therefore, the session was a challenge for me to change my style and deliver something of maximum value having been given the luxury of so much time. Using a practical workshop session adapted from the information gained during my

MBTI certification process, I was able to slowly take the athletes and coaches through a self-typing process. This journey through the personality types was carried out in such a way that all present were able to start appreciating personality similarities and differences along the way by actually seeing the proximity of themselves and colleagues on different personality factors. Each athlete and coach had their own reference material to work with throughout the session to help determine and record their own specific personality preferences. The session took place within a large physiology laboratory, where there was ample space to move around. The physical space was important because I needed to be able to physically map out the personality differences by gradually creating a personality type table on the floor using ropes to create the different sections.

Actually being able to see different personality preferences standing in different areas of the room made a significant impact on understanding tensions and harmonies that existed within the squad. The practicality of the session also meant that differences in personality preferences could be explored as being complementary approaches, rather than sources of conflict. Furthermore, it was possible to have a non-threatening, often humorous, approach to exploring different personality preferences within the squad, with a view to challenging the squad to consider how they could become more skilful at using the knowledge they were gaining to ensure they were maximizing the quality of working relationships, rather than simply fulfilling their personality stereotype. Therefore, athletes and coaches were challenged to understand the natural personality strengths that they bought to a crew/group, but more importantly, to understand how they would need to be skilful in extending their preferred behaviours and 'working out of preference' for the good of a collective unit. Exploiting existing personality strengths and becoming skilful at flexing personal approaches with different colleagues enabled the squad to see if this approach would help them become better athletes, better crews and more prepared to take on the rest of the world.

In addition to the generic introduction around personality type, the MBTI profile was used to help everyone identify how personality type becomes exaggerated under pressure. This allowed situation-specific self-awareness and awareness of crew mates to be considered. Importantly, with the raised knowledge of specific reactions under pressure the athletes could prepare for the pressure reactions to ensure that a negative impact did not result on other athletes at crucial times. All of this work laid a very positive foundation for specific crew work later in the season and in ensuing seasons. Many discussions had taken place before around managing different reactions to pressure within crews. However, the specific framework of the MBTI provided some key reference points that the athletes could use to help understand the differences and also predict the likely responses.

On completion of the practical session, the athletes and coaches had a rudimentary understanding of their own personality type as well as an appreciation for the personality preferences of key colleagues. To further support the learning, each athlete was then provided with a one-side summary of their personality type, according to the MBTI theory. The athletes and coaches were encouraged to read the profile to check if they felt it reflected how they would begin to describe themselves. The athletes and coaches were also encouraged to share some of their learning from the exercise with other members of their crew, with a view to determining how best to get the most out of each other at key times in the season.

As a final point of note for the athletes and coaches, they were challenged to ensure that the new information was not used in a negative manner. Specifically, the following provisos were suggested:

- Do not hide behind your 'type' – 'I'm sorry, I can't do that, I'm the wrong personality type'.

- Do not simply try to 'type' other people and make assumptions about how they are and how they are going to be.

- Do not let your 'type' allow self-fulfilling prophecies to happen – 'I couldn't help that, it's just the way I am'.

Therefore, the coaches and athletes were specifically challenged to:

- Practise flexing your approach for those kinds of situations when you know your natural preference does not lead to the best results.

- Work out where your 'blind spots' (i.e. things that you don't know about yourself that others do know) might be and plan how you might compensate for them.

- Practise understanding why other people might be seeing things differently from you and see if you can adapt the way you interact with them to get on the same page.

- Practise identifying where your differences in personality actually create a complementary set of skills, rather than an inevitable source of conflict. Ensure you are all benefiting from the different strengths you bring to the crew/squad.

From general concepts to specific applications

Having carried out the general introduction to the MBTI framework, many avenues were opened for specific applications of the output. The personality type information became increasingly important for informing the application of sport psychology concepts in an individualized manner. Concentration cues, confidence-building strategies, feedback requirements and goal setting approaches were all made more relevant to individuals as a result of the knowledge gained from the typing process. Rather than taking a one-size-fits-all approach to mental skills, it was possible to more effectively tailor the approach with which concepts were introduced and how they were structured, or help an athlete appreciate why they were struggling to implement a particular skill in certain situations. This has created a much more mindful application of tools and technique, which ultimately has increased my confidence as a practitioner.

Equally, communication preferences, learning styles, leadership approaches, response to change and stressors within groups can be more readily understood and exploited through the MBTI framework. My ability to present, and match, interventions effectively has been enhanced as a result of utilizing basic MBTI theory and knowledge. Having the ability to talk openly about these core elements of human interaction and challenging athletes and coaches to improve their 'skill' in understanding themselves and interacting with each other, positions the role of psychologist far beyond merely helping administer key psychological skills training.

For example, with one particular crew, the MBTI profiling has been invaluable for me as a practitioner and also a point of continuity for athletes and coaches with ongoing work over the last three years. For this particular crew, there has been gradual turnover within personnel over the three-year period, and with each change there has been a subtle shift in crew personality dynamic. As crew changes have happened, there has been a quick recognition of the personality preferences within the crew and a plan of action for how the crew need to manage them accordingly. Different combinations of athletes have resulted in different potential sources of conflict, harmony, blind spots and strengths, and all of these have had to be consistently referred to and exploited, or avoided accordingly. Having a framework for quickly capturing this information has meant that the crew have been able to take on collective aims for delivering the shared 'out of preference' behaviours, as well as identifying leadership roles within the crew as a function of fitness for purpose from a personality perspective.

Identification of this crew-specific context means that the crew have a competitive reason for ensuring that they become as good as possible at exploiting their psychological talent, experience and knowledge; the more they understand themselves individually and collectively, the more they will benefit in performance terms because they will be interacting as effectively as possible, as often as possible. Therefore, the crew has to work hard on gaining maximum benefit from the positives of shared thinking when similar preferences exist, and even harder on ensuring the shared blind spots do not undermine the ability to truly fulfil collective potential.

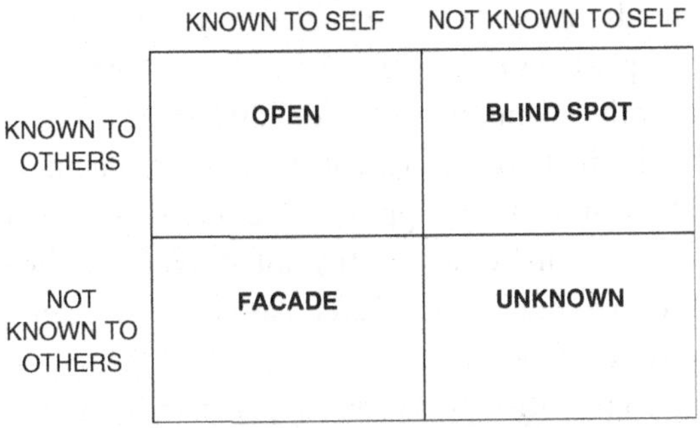

	KNOWN TO SELF	NOT KNOWN TO SELF
KNOWN TO OTHERS	OPEN	BLIND SPOT
NOT KNOWN TO OTHERS	FACADE	UNKNOWN

Figure 7.3   The Johari Window.

Creating a shared performance identity

The Johari Window and Competitive Knowledge Window

The second activity under consideration is the adaptation of the Johari Window (Luft and Ingham, 1955, see Figure 7.3) to help consolidate the knowledge that a crew has developed about itself into a performance-focused identity. In turn, this performance identity helps the crew make decisions about their optimum approach to a sequence of races at a regatta or in a specific one-off race.

In its own right, the Johari Window is a very useful exercise to carry out within groups to help raise all-round awareness levels. The four-frame 'window' is used to help individuals understand how they are perceived as well as to help others have an increased understanding of them as an individual. Within a group setting, the group would normally work through guided discussions or activities to create a clear picture of what is currently in the 'open arena' in terms of skills, attitudes, behaviours, feelings, motivations or experiences. The open arena refers specifically to everything that is clearly on view to everyone, in the public or team domain. Therefore, both the individual and other members of the group would identify what they believe to be known both by the individual and by other members of the group. With the open arena established, it is possible to provide feedback to a person in the blind spot area. The blind spot area refers to information that the group knows about the individual, but is not known by the individual about themselves. For example, the group may very clearly know early signs that an individual is feeling under pressure because of their behavioural changes, and making the individual aware of this allows them to benefit from the external knowledge, thus increasing the open-arena size. To further increase the size of the open arena it is possible for an individual within a group to disclose thoughts, feelings, attitudes and behaviours to the wider group that they believe are unknown to the group. Such personal disclosure decreases the size of the 'façade' or hidden-self, and ensures that the group is more aware of key knowledge that the individual holds themselves. The final area is that of the unknown or potential. Within a group, it is possible for there to be undiscovered elements to a person for all concerned; neither the group, nor the individual themselves has unearthed certain talents, abilities or attitudes that may actually be of great benefit to the group. Opening up discussions that consider what opportunities exist to stretch the individual and ultimately unearth talents can be very motivational. Working through each of the four areas, with the aim of making the open-arena as large and relevant as possible, is the clear aim of this exercise.

However, from a competitive performance perspective, it was clear to me that the framework provides greater benefit if it is adapted to capture known information about 'ourselves' vs the opposition. It has been noted that shared mental models are effectively a collated knowledge base that team members all hold around the task, each other, goals and strategies (Cannon-Bowers, Salas and Converse, 1993; Klimoski and Mohammed, 1994). Creating a process that allowed formal sharing of this kind of information, with a view to utilizing that knowledge to further refine strategy, would maximize the chances of a crew being greater than the sum of its individual parts. Therefore, an adaptation of the Johari Window, as seen in Figure 7.4, was created to form the Competitive Knowledge Window.

Using the reference points of 'us' and 'opposition' it is possible to populate the new window with simple but critical information based upon what is known and unknown at any point in time. Racing opportunities on the international rowing stage

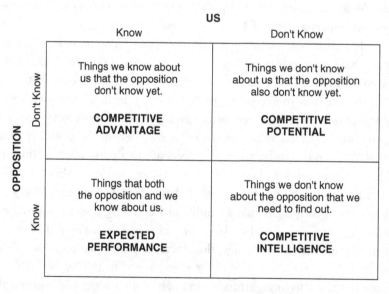

Figure 7.4   The Competition Knowledge Window.

		US	
		Know	Don't know
OPPOSITION	Don't know	Professionalism. Pride. Desire to win, passion. Quality of recovery ability. Keeping the big picture goal in mind – peaking. Faster first 500 m The quality of the winter training and what it's done for us physiologically. Common picture getting stronger. Our performance has moved on since Munich (2 x 1000 m). We can push the boundaries now with conviction. Getting closer to having a 'complete' 2k race.	Last 500 m speed. How fast we can go in the first 500 m. We don't know how quickly we're progressing relative to the opposition. How fast we can ultimately go. The impact of the changes we've made on our performance.
	Know	Good mid race pace. We can win. Third 500 m step-up. Our line–up. Second half is our strongest half. The history of us winning is getting ever longer – we have a winning legacy.	Opposition's strategies/race plan.

Figure 7.5 A completed Competitive Knowledge Window.

are limited, so every interaction between crews is critical self-learning but also a key opportunity to influence perceptions held by others and knowledge gleaned about others. Very simply, the crew, including the coach, spend as much time as is required to fill in the grid, sharing views and knowledge, and agreeing the most relevant content to be included. An example of a completed Competitive Knowledge Window can be seen in Figure 7.5

It is important to note that the content of the Competitive Knowledge Window represents the agreed output that has meaning for all crew members. The content does not simply represent every idea that was put forward. In the top left-hand corner, the aim is to identify those areas of self-knowledge that constitute 'competitive advantage'. These are typically strengths, tactics, technical changes, technological refinements, training performances and performance breakthroughs known to the crew that have not as yet been experienced by the opposition. Understanding when to implement these in competition and what the expected impact of them upon performance will be, is an important part of the conversation so that the crew collectively agree when and why competitive advantage is being revealed to the opposition. In the worked example it is important to note that all members of the crew and coach contributed to the elements in the competitive advantage section, resulting in a pooling of perceptions within the crew. Throughout the exercise, the athletes and coaches were formally sharing their consolidated views of everything that had taken place in recent weeks, and as such were forming an ever greater understanding of crew identity.

It is also possible to see from the worked example that the competitive advantage components come from core performance evidence, in-performance elements, preparation elements, recovery elements and attitudinal elements. The range of influences upon competitive identity ensures that the robustness of the resulting profile is maximized.

Next, the crew consider those things that they still do not know about themselves. 'Competitive potential' comprises those factors that are still to be developed and represent an opportunity to get excited about what might still be achieved and what the impact of changes/improvements might ultimately add up to in the future. Here, the crew can keep an eye on the future and maintain a focus on the key things that will help them remain motivated to keep improving and keep training with as much quality as possible. In this area of the window, it is possible that racing will allow some things to become known, given the opportunity to test performance elements out in the heat of competition. This area of the window can lead the crew to a situation where they can collectively pose questions to which they are actively going to try to find definitive answers in competition. For example, in Figure 7.5, the components related to first and last 500 m speed might result in some specific race strategy decisions to focus on these areas in isolation, rather than simply focusing on the outcome of the race. Committing to a race process as more important than the overall result obviously requires 'buy-in' from everyone, so the discussions around how to answer the questions ensure that every member of the crew has the same picture of success for each race. Given different goal orientations that will inevitably exist within crews, this explicit and constant affirmation of goals that are right for 'us' is key.

The third component to be completed is the 'expected performance' quadrant, which draws upon knowledge that is already in the public domain about the crew, the received wisdom. This open area of knowledge makes it very simple to consider what the opposition are going to be aware of about the crew and how their perceptions or psychology may be influenced (positively or negatively) as a result of what has gone in the past when the crews have been in competition. We have often used this window to reinforce the importance of crew self-presentation at regattas, both on and off the water, in training and racing, to help ensure that any preconceptions held by the opposition that are helpful to winning psychological 'battles' are being perpetuated. This quadrant is a very helpful confidence reinforcing area and also helps to ensure that received wisdom is exploited effectively when it comes to competing. However, the 'competitive potential' quadrant is also used to ensure that received wisdom does not become a limiting factor in terms of 'our' performance, because the crew are always willing to learn from what has gone before, but also completely willing to re-write the rule book when it comes to what should and should not be possible in racing. The crew are also actively thinking about how to influence the contents of the expected performance quadrant in the future by delivering performances that create as much received wisdom as possible that leads the opposition to conclude that they are unlikely to win.

The final quadrant to complete is the 'competitive intelligence' quadrant, which simply identifies any specific knowledge that the crew wants to target about any of the opposition. Coaches acquire a lot of video and technical information about other crews at regattas, so the crew can decide what part they want competitive intelligence to play in fuelling their overall racing psychology. Within this quadrant it is important to ensure that any information that is entered adds to the racing knowledge and performance of the crew. It is very easy to gather a whole load of information about the opposition that actually does nothing to further the racing psychology of a crew, but simply makes them overly informed about the opposition. In the example in Figure 7.5 it is clear that on this occasion of completing the exercise very little attention was being paid to the need to focus on the opposition. This particular example was carried out before the first race of the competitive season. Therefore, the crew were focused solely on controlling their performance and delivering that with maximum effectiveness, which would then allow conclusions to be drawn about the opposition relative to their own performance.

The whole Competitive Knowledge Window exercise is a very worthwhile communication tool and can be used as often as desired throughout a season as knowledge changes occur. With increased usage the crew becomes more aligned more quickly and the exercise is completed more efficiently. This increased speed of achieving consensus is also very helpful from a crew psychology point of view when it is clear to the athletes and coach that they are sharing the same thinking more of the time. From an overall philosophical approach, it is useful for the crew to be considering key questions around how they move information from the 'competitive potential' quadrant into the 'competitive advantage' quadrant as effectively as possible, before finally showing their hand and turning those qualities into 'expected performance'.

On a practical level, the Competitive Knowledge Window exercise has proved to be very powerful in terms of reinforcing fundamentals of racing psychology as well as helping to build confidence. Critically, the exercise helps to exploit the knowledge resource of the crew and reinforces the importance of the identity of the crew. Just as a great company may take great confidence and identity from its trade secrets, so too the crew begins to increase togetherness as a result of formally identifying their unique knowledge pool and determining how they will exploit and protect this knowledge. At the highest level, these subtle processes can make critical differences during the moments of pressure, so they are worthwhile investing time in to ensure that any advantage is gained.

7.4 Evaluation of interventions

The evaluation of the activities reported took place within the ongoing monitoring and evaluation of the sport psychology provision. As outlined previously, I consider the work carried out to be part of an ongoing action-learning project. Therefore, as a matter of course, there are a number of processes engaged in that help determine the immediate and ongoing value of work that has been carried out to determine if a 'problem' has been solved. Weinberg and Williams (2006) note that evaluation should be a continuous process for integrating and implementing psychological skills training programmes, and this is even more the case when considering the broader provision of psychology support beyond psychological skills training. Therefore, as with any needs analysis, evaluations are formed primarily through qualitatively based collection of information. The following are sources of information and reference that are used to determine activity value and future action:

- Discussions with the coach regarding the perceived value of any session to their ongoing planning and coaching.

- Discussion with key athletes regarding the applied relevance of the work that has been carried out.

- Willingness of the coaches and athletes to take ownership of activities and drive further applications of the concepts by actively engaging the psychologist.

- Personal reflection upon the value of the session delivered and the receptivity and engagement of those involved.

- Annual review by coaches, athletes and sport psychologist as part of wider perfor-mance review of the entire sport.

I have also been influenced in my applied work by the social validation techniques associated with single-case research designs (Kazdin, 1977). By using the notions of social comparison (comparing athletes and coaches who have taken part in an activity with other members of the squad who have not) and subjective evaluation (direct feedback from the recipient of value or feedback from others associated with the recipient), it is possible to use ongoing observations within the sport to pick up signs of value and impact. I also elicit specific feedback from coaches and athletes about the perceived value of work that has been carried out.

Much of the emphasis of my consultancy work focuses on competitive reviewing and learning (gaining competitive advantage by reviewing and learning with greater speed and effectiveness than opponents), which encourages the athletes to be constantly identifying the impact of choices made and actions taken. Additionally, there is a constant focus on identifying the next high impact action to take as a function of what has been learned. Therefore, the psychology work that is carried out is reviewed with all other areas of performance that are being developed. This results in the athletes evaluating if what they are doing is 'making the boat go faster' and, if it is not, then refinements are sought. Indeed, in terms of evaluation of my impact, I am often asked to identify how I believe I have contributed to making boats go faster, which is perhaps the ultimate evaluation of impact.

As well as this broad evaluation of the work, some specific results of the two activities were also evident. In relation to the MBTI work there were immediate results of raised self-awareness and awareness of others within the squad. Athletes and coaches started to have conversations that focused more specifically on different learning and coaching styles to optimize feedback and skill development. Athletes were able to more readily identify the reasons behind themselves and coaches differing on the relative importance of different elements of training, and as a result could make efforts to work in a more complementary manner. Most importantly, athletes and coaches were able to start re-framing personality differences as a potential barrier to fulfilling potential. With this shared view it was then possible to work collaboratively to ensure that personality did not 'get in the way' of performance. With a frame of reference and shared set of principles to work with, it made my work a lot easier in terms of facilitating optimal work relationships and constantly challenging everyone to be psychologically the best performer they could be.

In relation to the Competitive Knowledge Window there were less specific outputs. However, the crew in question were able to go into their first regatta with a confidence and a very clear way of delivering their own performance. Collectively 'grounding' the confidence to the same sources was important so the crew and coach could be absolutely on the 'same page' when finalizing the race plans to be executed. Having clarity over shared process goals, especially in the early season, makes a critical impact upon the performance that is delivered and then the quality of training focus that is achieved after the regatta. Assuming shared process goals can instantly create a situation where all crew members are not quite aligned in their thinking, so the only conclusions that can be drawn relate to 'how we race when we are not all quite focused on the same goals'. Importantly, with two or three races at a regatta, the crew used the exercise after races to ensure that they were learning from each performance. Being able to test out whether specific bits of predicted knowledge were affirmed, or needed more intelligence gathering, was a useful 'in-performance' application of the exercise. The willingness of the crew to use the window as a reviewing tool within regattas demonstrated a perceived benefit to having the framework.

7.5 Evaluation of consultant effectiveness/reflective practice

My personal evaluation will always revolve around key questions of:

- Was I prepared for the specific nature and focus of the session?

- Was I fully focused throughout the session?

- Did I demonstrate confidence and conviction in my knowledge and ideas?

- Did I ask great questions and give great answers in equal proportion?

- Did the session result in concrete actions or recommendations for the coach, athlete or both?

- Did everyone contribute and have the opportunity to understand the session for maximum personal benefit?

- Was the time used effectively, given that it is a precious commodity?

- Did I get session notes through to all concerned in sufficient time to build on the momentum from the session?

- Is there anything I need to reinforce after the session through my communications that was either omitted or is a key point for future sessions?

- Are the coaches and athletes showing signs of taking ownership of key concepts that I have delivered?

My many years of work within one sport have allowed a considerable amount of reflective practice. Accordingly I have some general reflections, which are important because they have influenced the philosophy behind the work reported in this chapter. 1 also have some specific reflections about the two activities reported, which will be addressed first.

Specific reflections

Having focused on the utilization of the MBTI within a sport setting, there are some useful observations and personal learning. First, the session was positively received and created an immediate impact for the athletes and coaches, simply through raising awareness. Much of the positive nature of the impact was a direct result of the fact that the session was not traditional mental skills training, but was focused upon interpersonal relationships. The positive attitudes towards a non-performance-focused session reinforced some growing ideas regarding the positioning of sport psychology for athletes and coaches. Allowing the athletes and coaches to receive structured information that stimulated their thinking around psychology generally did a very good job of promoting the importance of all things psychological. Helping people become curious about their thoughts, feelings and actions provided a much firmer base upon which to introduce recommended psychological development. I was able to provide a strong rationale for why particular approaches to in-race thinking, confidence building, goal setting or management of pressure were being recommended and as a result a much greater receptivity towards these areas of performance psychology was achieved. This enabled a much greater individualization, an obvious factor that is important to the adherence to the psychological recommendations (Shambrook and Bull, 1999). Simply presenting psychological skills training as 'something all good athletes should be great at', does not typically result in such levels of receptivity. Since the original work into adherence to psychological skills training (Bull, 1991) and my own Ph.D., it has become increasingly clear to me that it is critical to ensure that a general grounding and interest in broader psychology is developed prior to the introduction of specific psychological skills training. Ensuring such a grounding in psychology generally and interest raising for the topic helped to move the athletes and coaches through the Stages of Change (Prochaska and DiClemente, 1983) and ultimately maximized the athletes' desire to invest time and energy in enhancing their 'mental fitness' or mental skill level.

From an adherence point of view it is also likely that the particular way in which a consultant introduces an athlete to a programme of work has a result in their commitment to it. It has been well documented that receptivity alone is not enough to guarantee implementation (Blinde and Tierney, 1990); however, receptivity can be maximized by presenting information in a way that is most likely to appeal to specific personality types. Furthermore, once receptivity has been optimized, then programmes can be structured in a way that maximizes the likelihood of consistent implementation being carried out. Such a thought process leads to the notion that practitioners should introduce their work to different athletes in the most appropriate way to match individual preferences. At the elite level, such individualization of programme introduction and structuring would maximize speed of impact upon performance.

As stated, the main aim of using the MBTI was to raise self-awareness and awareness of others. This has been essential in creating a performance-focused approach to understanding and exploiting personality. The concepts of self-awareness and awareness of others are also core features that underpin emotional intelligence (Goleman,

1995), a core set of competencies for any team or individual. Being able to challenge individuals and teams to be as 'emotionally intelligent' as possible is a useful way of positioning this kind of work. The rowers are typically used to appraising themselves and each other in relation to physical and technical qualities. Bringing in an approach and language that has allowed personality qualities to be understood and used in a similar, performance-focused manner, has added to the mix of factors that explicitly contribute to the success of the crew.

I am sure that other personality frameworks would provide an equally useful foundation for such psychology work, so this is not an MBTI-specific promotion. Importantly though, this approach has been enduringly successful because it has consistently been focused on ensuring that any conversations around personality are aimed at improved rowing performance, through an accepted and understood set of principles. Without demonstrating how the knowledge actually helped boats move faster through the water, the impact would not have been so great. Thus ensuring that personality principles are connected to final performance was an important point to constantly apply to the work.

The application of MBTI profiling has been a useful example of how my approach to delivery in the role of sport psychologist has changed over the past 10 years. Originally, I was focused on making sure that I delivered the approaches that the sport psychology literature had highlighted to me *should* be of value to coaches and athletes, because the logic behind the interventions or ideas was plain to see for the sport psychologist and should therefore be embraced by the consumer. For the most part, I would describe this work as being focused on those topics and approaches of greatest interest to psychologists when examining the overall psychology of sport. It is now clear to me that the things of greatest interest to psychologists are not necessarily of interest to coaches and athletes. I am now much more focused on the psychology of delivering results and accordingly the approaches that come primarily from understanding explicitly what is of greatest value to athletes and coaches who are constantly focused on achieving results. This has led to me looking to diverse sources of inspiration for delivering psychology related support (e.g. Buckingham and Clifton; 2001; Collins, 2001; Covey, 1989; Goleman, 1995; Robbins and Finley, 1996). These popular psychology books have provided excellent concepts and approaches that can be readily packaged into very accessible psychology education. They also demonstrate consistently that it is essential to make psychological concepts easy to understand, able to be owned by anyone and nicely packaged into simple, memorable, readily applicable packages. This may end up looking like 'tricks and gimmicks' from the view point of the academic world, but is an essential part of ensuring the maximum impact of psychological work.

This openness to a greater diversity of psychology literature also reflects a shift in my professional paranoia. I am no longer driven by the thought of 'can I justify this approach theoretically?' Although that reference is still professionally important, I am more driven by a thought that revolves around 'have I delivered maximum impact for the athletes and coaches?' From a development perspective, this means that I now believe firmly in the importance of training practitioners in sport psychology to be accountable to both professional standards and delivering results-focused work, in equal proportion. Without this dual approach, I believe it is easy to end up in a situation where excellent sport psychology interventions get delivered from a theoretical perspective *only*, but in reality they do not achieve optimal impact for the coaches and athletes. I have certainly found information regarding evidence-based practice in a variety of health care settings (Trinder and Reynolds, 2000) particularly useful in helping me to manage more effectively the combination of evidence that is derived directly from professional practice (practice-based evidence) and that which is derived from existing theory and research. Being able to have equal confidence in the value of these sources of evidence opens up a greater flexibility in my thinking as a practitioner when it comes to determining how I might be most effective in any given situation.

The development of the Competitive Knowledge Window was a case in point regarding my shift in approach. The approach delivered, very simply, information that coach and crew could make use of in a number of ways when preparing for competition. With this particular approach, the success of it was demonstrated by two clear pieces of evidence. First, a further regular process was developed by one specific crew to ensure that the exercise produced clear outputs upon which the crew would focus its energies for a given regatta or race. This was an exercise of identifying key questions that needed to be answerable as a result of having raced, and ahead of every regatta the list of questions was produced and then systematically evaluated post-regatta, regardless of my presence. Second, the coach and the crew drove the timing of additional usage of the Competitive Knowledge Window, using it when they felt it most appropriate, rather than waiting for me to tell them when they should use it. When practices are being carried out independent of the psychologist, it is clear that a worthwhile activity has been delivered. I have no desire to make myself redundant as a practitioner, for a whole host of reasons, but I do have a strong desire to introduce psychological work that does not require me to be present for it to be implemented.

General reflections upon practice

First, there are the general reflections that now underpin my approach. In terms of my mindset as a sport psychologist, I am totally focused upon sport psychology being a positive, value-add service. I am increasingly focused on the positive psychology focus offered by Seligman and Csikszentmihalyi (2000) and I have focused these ideas to fit further into the competitive ethos of elite sport, ensuring that positive, desirable characteristics are actively pursued in order to gain a competitive advantage. Too often sport psychology still gets spoken about as a problem-fixing service. You will see that both of the concepts within this case study have been delivered to maximize psychological resources and to gain competitive advantage, rather than as a reaction to a problem that needed fixing. The language I have used has also steered away from 'intervention', 'problem-solving' and terms that may have their legacy routed within clinical models of remedial psychology provision. I believe that sport psychology used simply as a problem-solving agent has limited scope within elite sport and such an approach does a disservice to the potential impact of psychological preparation upon performance. Athletes do not wait to become physically unfit before they start putting the physical work in. Sport psychology should be treated with exactly the same philosophy at the highest level, with athletes and coaches constantly striving to be as 'mentally fit' as possible, rather than waiting for psychological weaknesses or issues to transpire before considering using sport psychology.

In a related point, I believe that with performance-focused psychology it should always be possible to have open conversations about the focus of performance-enhancing psychology work. Confidentiality issues can sometimes get in the way of fully trusting, collaborative relationships between psychologist, coach and athlete. I believe that it is important that there are shared conversations about the focus of psychology work. A physiologist would seldom carry out physiological monitoring with an athlete and then not share the results and recommendations openly with both coach and athlete. Failure to do this would result in there not being a united plan of how to exploit training to enhance the physiological profile to greatest effect. If any kind of performance is to be improved, then all parties must understand the current requirements and the suggested forward plan. Performance-focused psychology conversations can take place in exactly the same open manner, with a view to taking mental fitness to the next level, so my default position when consulting is that conversations will be shared, unless there is an explicit request not to from the athlete or coach. A clear delineation between performance-enhancing work and work that requires strict client confidentiality is of course essential. This requires time to build the trust that both roles can be fulfilled with equal positive intent and effectiveness.

It has also become clear to me that being a psychologist is very helpful to my own performance. As Hardy *et al.* (1996) point out, many practitioners end up benefiting themselves and make consistent use of the skills they teach to others. Being a psychologist does not make me immune to the influences of psychology upon my own performance. Identifying the psychological challenges faced by sport psychologists is worthy of discussion so that we can prepare all practitioners for the psychological demands they will face, as well as with technical expertise. There is more to sport psychology than the theoretical and scientific underpinnings and I believe that increasing the number of open conversations about what it takes to actually deliver sport psychology within high-pressure, high-expectation, performance contexts is an essential element in pushing applied sport psychology delivery to the next level. My opinion is that there has been little shift in relation to the delivery of sport psychology in recent years and there have been few compelling breakthroughs in research terms that have led to a wholesale change in what is delivered to coaches and athletes. Therefore, if gains are to be made, they will come in understanding the art of delivery more exactingly, so that we can prepare more effectively as a profession to rise to the challenge and make a difference when asked.

I have felt the need to manage my own expectations, confidence levels and motivation on a regular basis primarily as a function of being a part-time member of the team, but wanting to have a 'full-time' impact in my work as the psychologist. Given sporadic involvement throughout a season, it is often difficult to progress ideas, relationships and applied work sufficiently quickly to always feel like my expertise is producing maximum benefit for all concerned. Maximizing impact upon thought processes, attitudes, beliefs and psychological skill application through 'brief' forays into the applied arena is a challenge for the majority of practitioners and far from an ideal way to influence these key elements of psychology. Once again this means it is critical that we, as a profession, find ways to consult with and through, full-time members of the support team (coaches and science/medicine staff), as well as finding ways in which psychology can be truly integrated into the day-to-day training programmes of athletes.

The role of sport psychologist for the GB Rowing team has been highly rewarding and has constantly challenged me to find ways of 'raising my game' to pass on psychology consultancy that will make a difference in performance terms. By having the confidence to think beyond psychological skills training and think more about helping to develop working knowledge of all things psychological within cutting edge sport, I have been able to develop new skills and techniques that integrate psychological development into performance discussions on a regular basis. My performance as a psychologist keeps improving, just as the performance of the athletes and coaches improves. To not challenge myself to 'get better' would be to isolate myself from the expectations that are placed on everyone within the sport. Even after 12 years of work within the sport, there are still areas in which it is necessary to improve the acceptance and utilization of sport psychology. However, the successes achieved with the specific approaches that are outlined in this chapter fuel the motivation to keep challenging myself and those with whom I work to find the most effective ways for the whole squad to become the mentally fittest performers in world sport.

7.6 Summary

This chapter has focused primarily on two pieces of applied sport psychology work with the GB Rowing team that go beyond psychological skills training. Psychological skills training is an essential part of any sport psychology provision. However, the focus on raising collective awareness about different psychologies within a squad environment demonstrated the importance of wider psychology education, as part of ongoing work within a given context.

The MBTI was used effectively to raise awareness of different ways of thinking and working within a squad context. However, from a performance-focused perspective it is important to note that this work constantly emphasized the importance of not simply knowing about personality, but making use of this knowledge to bring greater collaborative work towards delivering performance. The outcome-focused emphasis was an important part of the application of this activity to the elite environment in question.

The Competitive Knowledge Window was used to demonstrate the importance of being able to take concepts from wider areas of psychology to help improve performance in areas such as shared decision-making and collective identity. This activity helped to ensure a crew mindset was being explicitly created. Importantly, this activity built the foundation for competitive learning for the crew in question and as they took ownership of this simple, but effective, shared thinking tool, they were able to further develop it to meet their racing and performance needs.

Questions for students

1 What do you understand by the term 'action learning' and how it might apply to an ongoing consultancy role within a sport?

2 By raising self-awareness and awareness of others through discussions of personality differences, how can this help to make a difference to training and competition effectiveness?

3 How would the Competitive Knowledge Window be useful for a newly formed team compared with a well-established team?

4 Discuss ways in which you think psychologists can maximize the chances of athletes and coaches taking ownership of activities that will improve mental fitness.

5 What kind of mental skills do you think are required for a psychologist who delivers within elite sport?

References

Blinde, E.M. and Tierney, J.E. (1990) Diffusion of sport psychology into elite U.S. swimming programs. *The Sport Psychologist* **4**, 130–144.

Briggs-Myers, I., McCauley, M.H., Quenk, N.L. *et al.* (1998) *MBTI Manual: A Guide to the Development and Use of the Myers-Briggs Type Indicator* (3rd edn). Consulting Psychologists Press, Palo Alto, CA.

Buckingham, M. and Clifton D.O. (2001) *Now Discover Your Strengths*. Simon and Schuster, New York.

Bull, S.J. (1991) Personal and situational influences on adherence to mental skills training. *Journal of Sport and Exercise Psychology* **13**, 121–132.

Cannon-Bowers, J.A., Salas, E. and Converse, S. (1993) Shared mental models in expert team decision making. In: Castellan, N.J. (ed.), *Individual and Group Decision Making*, pp. 221–246. Lawrence Erlbaum, Hillsdale, NJ.

Collins, J. (2001) *Good to Great*. HarperCollins, New York.

Covey, S. (1989) *The Seven Habits of Highly Effective People: Powerful Lessons in Personal Change*. Simon and Schuster, London.

Goleman, D. (1995) *Emotional Intelligence*. Bantam Books, New York.

Hardy, L., Jones, G. and Gould, D. (1996) *Understanding Psychological Preparation for Sport: Theory and Practice of Elite Performers*. John Wiley & Sons, Chichester.

Kazdin, A. E. (1977). Assessing the clinical or applied importance of behavior change through social validation. *Behavior Modification* **1**, 427–452.

Klimoski, R. and Mohammed, S. (1994). Team mental model: construct or metaphor? *Journal of Management* **20**, 403–437.

Luft, J. and Ingham, H. (1955). The Johari Window: a graphic model of interpersonal awareness. *Proceedings of the Western Training Laboratory in Group Development*. UCLA, Los Angeles, CA.

Myers, I.B. (1962) *Manual: the Myers-Briggs Type Indicator*. Educational Testing Service, Princeton, NJ.

Prochaska, J. and DiClemente, C. (1983) Stages and process of self-change in smoking: toward an integrative model of change. *Journal of Consulting and Clinical Psychology* **51**, 390–395.

Robbins, H. and Finley, M. (1996) *Why Teams Don't Work: What Went Wrong and How to Make it Right*. Orion Business, London.

Rothwell, W.J. (1999) *The Action Learning Guide Book: A Real Time Strategy for Problem Solving, Training Design, and Employee Development*. Jossey-Bass/Pfeiffer, San Francisco, CA.

Seligman, M. and Csikszentmihalyi, M. (2000). Positive psychology: an introduction. *American Psychologist* **55**, 5–14.

Shambrook, C.J. and Bull, S.J. (1999) Adherence to psychological preparation in sport. In: Bull, S.J. (ed.), *Adherence Issues in Sport and Exercise*, pp. 169–196. John Wiley & Sons, Chichester.

Thomas, P. (1990) *An Overview of the Performance Enhancement Process in Applied Psychology*. US Olympic Center, Colorado Springs.

Trinder, L. and Reynolds, S. (2000) *Evidence-Based Practice: a Critical Appraisal*. Blackwell Science, Oxford.

Weinberg, R. and Williams, J. (2006). Integrating and implementing a psychological skills training program. In: Williams, J.M. (ed.), *Applied Sport Psychology: Personal Growth to Peak Performance*, pp. 425–457. McGraw-Hill, New York.

8

Delivering Educational Workshops for Age-Group Rugby League Players: Experiences of a Trainee Sport Psychologist

Jenny Page

University of Portsmouth, Portsmouth, UK

8.1 Introduction/background information

At the beginning of the work described in this chapter I was a year into the three year BASES Supervised Experience (SE) process (and therefore a Probationary Sport Scientist – Psychology Support). There had been limited opportunities to design and deliver workshops or interventions to sports performers. At this point in time, I had an undergraduate degree in sports science and a taught postgraduate degree in sport psychology, so my strengths were based on theoretical and research knowledge of sport psychology. It was identified through the SE process, in conjunction with my supervisor, that I needed to become involved in delivering workshops to groups of performers in order to improve my competency in applied consultancy skills. Fortunately, an experienced colleague, involved in providing physiological support to a professional rugby league club, invited me along to a fitness testing session where I met with the scholarship manager and discussed the possible implementation of sport

Applied Sport Psychology Edited by Brian Hemmings and Tim Holder
© 2009 John Wiley & Sons, Ltd

psychology workshops. The scholarship manager was enthusiastic and explained that they were responsible for four squads (under 16s, under 15s, under 14s and under 13s). Each squad had a head coach and trained on average twice a week, with approximately 15–20 players in each squad.

Prior to designing the workshops I was keen to develop my knowledge of rugby league in order to aid the development of credibility and trust (Poczwardowski, Sherman and Henschen, 1998). Bull (1995) stated that having technical knowledge of cricket enabled him to become integrated into the coaching team, facilitated acceptance with the players and helped to provide practical examples of mental training techniques. During this process, I observed some of the squads in action to obtain information regarding the demands of the sport. Having no previous experience of the sport I also read a beginner's guide produced in collaboration with the Rugby Football League (1994) entitled *Know the Game: Rugby League*, in order to understand the fundamentals of the game, the rules of play and the basic skills required.

This chapter will discuss the needs assessment, development and monitoring, and evaluation of an educational workshop programme that I delivered in the first year that I worked with the scholarship programme (season one). It will then address the needs assessment, development and monitoring, and evaluation of a workshop programme that I delivered the following year (season two). Particular emphasis will be placed on highlighting the issues that arose as a result of the first season's workshops, and how they influenced interventions delivered in the second season.

8.2 Initial needs assessment – season one

Between-method triangulation enables a balanced and holistic approach to the needs assessment (Arksey and Knight, 1999). However, due to time-constraints, I was granted limited access to staff and players prior to service delivery. Initial interaction and informal observations seemed to indicate little experience of sport psychology by the coaching staff. However, the scholarship manager was very supportive of what sport psychology might have to offer and was therefore contacted to enable an informal needs assessment to take place.

Initial meeting with scholarship manager

During the meeting the scholarship manager suggested that he was keen to have the workshops delivered between September and December in conjunction with the squad training sessions. He requested that the workshops be delivered to all squads from under 16s to under 13s, and that he would like sessions to be delivered on performance profiling (Butler and Hardy, 1992) and coping strategies (Bull, Albinson and Shambrook, 1996; Scully and Kremer, 1997). He also suggested that a session on goal setting (Locke, 1996; Locke and Latham, 1984, 1985) should be delivered to the younger players, based on his perceptions of what the squads needed and because he was already familiar with these techniques. Two other workshops were agreed upon, based upon my requests. These were an introduction to sport psychology and an evaluation session to conclude the workshops.

8.3 Intervention and monitoring – season one

Development and delivery of workshop programme

My experience as a trainee cognitive–behavioural practitioner informed the development of the workshops. All of the workshops were designed and discussed with my supervisor and the scholarship manager prior to delivery of the first workshop. Initially, I listed a title for each workshop and designed the workshop content. The material I had prepared to cover was significantly reduced after collaboration with my SE supervisor.

The workshops were all delivered in a small classroom, immediately after evening training sessions. All squads, along with their respective head coaches, were expected to attend each of the five workshops based on the request of the scholarship manager. There was an approximate three week gap between each of the workshops. A brief overview of the workshop programme is presented in Table 8.1.

The first workshop, entitled 'Introduction to sport psychology', was designed to enable me to introduce myself, and to introduce the concept of sport psychology to begin the important rapport-building process (Petitpas, 2000). During the workshop each player was asked to imagine their best and worst performances and then the consequences of positive and negative thinking were discussed. The younger group focused on the direct link between thinking and performance, whereas the older group were able to focus on both the emotional and performance consequences of different types of thinking. During this session the coaches were clustered at the back of the room, and did not contribute to the session.

Following this (and before workshop two), a coach discussion session was set up in order to prepare for the workshop on performance profiling (Butler and Hardy, 1992; Weston, 2008). During this session two coaching groups were formed, dividing the coaches of the older and younger athletes, and each group of coaches was asked to identify the 10 most important performance qualities for players in their respective age groups. This elicited some discussion between the coaches, who were unsure about how to prioritize the 10 most important qualities for their players. To try and resolve their concerns, we discussed how they managed to prioritize the content of their training sessions and then I tried to relate that to prioritizing the performance qualities based on the existing needs of the squads.

Table 8.1   Overview of season one workshops

Title	Delivery	Content	Activities employed
1. Introduction to sport psychology	Twice (under 13s and 14s) and (under 15s and 16s)	The link between thoughts and behaviours	40 minute mini-lecture and 20 minute group discussion of best and worst performances
2. Performance profiling	Twice (under 13s and 14s) and (under15s and 16s)	Filling in the pre-designed performance profile	20 minute mini-lecture followed by 40 minutes of the coaches defining the 10 qualities before the athletes rated themselves
3. Goal setting	Once (under 13s and 14s)	Setting SMART goals	45 minute mini-lectured followed by 15 minute interactive group discussions regarding goals that they would like to set
3. Coping strategies	Once (under 15s and 16s)	Centering and error parking	15 minute mini-lecture followed by practice of the techniques
4. Coping strategies	Twice (under 13s and 14s) and (under15s and 16s)	Control the controllables	40 minute mini-lecture followed by 20 minute interactive group discussions of what they were in control of
5. Performance evaluation	Twice (under 13s and 14s) and (under15s and 16s)	Effective performance evaluation	30 minute mini-lecture followed by 30 minutes evaluating their last game performance, and my performance in the workshops

During the performance profiling workshop (workshop two), I explained the potential uses of the performance profile to the players, which included recognizing their individual strengths and weaknesses as a platform for goal setting. After this, the coaches were asked to explain each of the 10 qualities that they had identified in the previous session and then give examples of what would constitute a rating from 1 to 10 to aid player understanding.

During workshop three the younger age groups were introduced to SMART (specific, measurable, adjustable, realistic, time-based) goal setting (e.g. Bull *et al.*, 1996). The acronym was explained to the players and this led to a lot of questions from the players as they were unsure how to determine whether a goal was specific enough. I felt ill-equipped to answer their questions at times and could not find a 'player friendly' definition of how to set a specific goal. After this we worked as a group through setting two goals, which seemed to help the players understand the notion of specific goals.

For the older players, workshop three involved an introduction to coping strategies. This workshop started by introducing centering (Scully and Kremer, 1997) and error parking (Bull *et al.*, 1996). The players were then given opportunities to practise both techniques whilst I worked with them through the processes involved to perform each technique. Finally the players were asked to identify times when they thought these techniques might be useful and share them with the rest of the group.

During workshop four on coping strategies, I asked the players to discuss all the factors that influenced their matches. They were then asked to discuss in groups whether they thought they were in control of each factor. This elicited some debate with regard to the control of certain aspects of their performance such as 'thinking'. The players were then asked to discuss whether each of the factors was helpful or unhelpful. From this, the players were then asked to concentrate on the controllable factors that were helpful to performance and discuss these within small groups.

The final workshop aimed to introduce the benefit of performance evaluation and involved the players evaluating past performances using a pre-designed performance evaluation sheet (based on Holder, 1997). Players were asked to focus on specific performance processes that they performed well and identify areas that they would like to improve as a result of the evaluations. At the end of the workshop players were asked to use a similar sheet to evaluate the workshop programme as a whole. They were asked to consider what they had learned and how it could be improved.

8.4 Evaluation of intervention – season one

Evaluations from players

The older players suggested that the practical elements such as centering and error parking were the most enjoyable parts of the workshops. However, some players had trouble acknowledging what they had learned; this suggested that they may not have engaged fully with the content of the sessions. When asked about what improvements could be made, many of the players suggested that the workshops needed to be 'less boring' and that they wanted to be training rather than 'stuck in a classroom'. When questioned in the final workshop, players also suggested that they did not know when they were going to use the psychological skills as they were unsure how they related to game situations.

Reflections – season one

In order to design and deliver effective interventions I utilized reflective practice to evaluate theoretical and evidence-based practice, evidence of the effectiveness of the intervention, professional practice of the sport psychologist, and ways to improve practice (Ghaye and Ghaye, 1998). Poczwardowski et al. (1998) suggest that, by paying attention to oneself, thoughtfully analysing consultations, and being aware of limitations, self-interests, prejudices and frustrations, practitioners will be in a better position to manage themselves and their practice effectively. Based on my reflections a number of issues were identified.

Reflection 1: lack of access to and rapport with support staff prior to the workshops

Prior to delivering the workshops, I had very little contact and communication with the head coaches and players of the specific squads that I would be working with. This was problematic since Anderson, Knowles and Gilbourne (2004) found that good communication was a key characteristic for effective consultants. Part of the reason for this was that there was only one month between meeting the scholarship manager and delivering the first workshop. The lack of communication and rapport with the coaches and players affected the workshops in a variety of ways. Firstly, the initial assessment was ill-informed, as I felt uncomfortable approaching the head coaches for help in designing the workshops as I had not previously had any contact with them. Through my inexperience, this led me to deliver workshops to which the head coaches had given little input. Being excluded from the design of sport psychology interventions has been shown to decrease the adherence of players (Maddux, 1993) and may also decrease autonomy, which can contribute to both extrinsic motivation and/or amotivation (Deci and Ryan, 1985; Ryan and Deci, 2000). As the coaches were not involved in the decision-making process, they did not seem to engage with the material, and as a consequence talked amongst themselves at the back of the room, particularly in workshop one.

Reflection 2: coaches' lack of psychology knowledge

Although the scholarship manager had a reasonable understanding of sport psychology, which enabled me to gain initial access, many of the coaches had not experienced sport psychology before. The lack of education regarding the potential benefits of sport psychology may have contributed to their lack of attention during the workshops. Pain and Harwood (2004) suggest that applied sport psychologists need to do more to demonstrate the value of their work to those within soccer and other sports. Although a brief attempt to highlight the potential benefits of sport psychology education and support was made in workshop one, it was clear that dedicating such a small amount of time to this very important issue did not aid the effectiveness of the workshops. On reflection, it seems that educating the coaches about the benefits of sport psychology was a very important part of the role of the sport psychologist (Williams and Kendall, 2007).

The coaches' lack of psychology knowledge was apparent throughout the coach education session. The coaches' opinion seemed to be that it was impossible to pick only 10 qualities for the performance profile and that it would therefore be of very limited use. Such direct criticism came as a surprise to me. I then had difficulty trying to 'sell' the benefits of performance profiling. Being aware of the benefits of psychology and being able to communicate these to others is therefore an important issue when delivering such workshops. Modifying the task difficulty may have been a useful technique, for example allowing the coaches to pick 20 qualities (in line with Butler and Hardy, 1992) may have been easier for the coaches, and may have generated more conversation and less criticism of the tool. However, I deemed that reviewing 20 qualities in the subsequent one-hour workshop would have been overwhelming for the young players. Additionally, the coaches appeared to find it very difficult in the performance profiling workshop to define what constituted a score of 10 for each quality. As a result the players seemed a little confused and perhaps did not fully engage in the process. This was a difficult situation as I did not want to interrupt the coaches throughout this part of the session. However, on reflection I felt that I should have checked the coaches' understanding of this process before delivering the session, or intervened to facilitate the session.

Reflection 3: being personable during workshops

The limited rapport with the players also influenced the delivery of workshop one (Introduction to sport psychology). After being asked to discuss their best ever performance, one of the under 15/16 groups reported that they could not remember their best performance. Rather than pursuing the issue and questioning further I moved on to the next player, who repeated that they too could not think of a best performance. After this point many of the other players copied, and I felt I was struggling to progress the session. At first I did not know how to deal with the situation and quietly started to panic; however after the fifth or six player replied with 'neither can I', I started laughing and made a joke regarding the memories of rugby league players. This seemed to promote conversation, and eventually the players were speaking more freely about their experiences. Being personable was identified by Anderson *et al.* (2004) as a key characteristic for effective consultants. Supporting this, Pain and Harwood (2004) found that Academy directors in football felt that sport psychologists must have the character to deal with the sport environment, the 'banter' of players and uncertain attitudes toward sport psychology.

Reflection 4: my lack of knowledge of the sport

Despite attempts to read 'know the game' books and several prior observations of training, my knowledge of the sport was still very limited when the workshops began. A key concern was that I would not use the right terminology and be exposed as someone who did not know the sport. Pain and Harwood (2004) suggest that wherever possible, examples given should relate directly to the experiences of athletes, using language appropriate to the sport. Being confident in your ability to do this is therefore an important consideration and although I was reasonably confident with the psychological content, the sport-specific terms were not as well learnt. This would suggest that preparation of the psychological and sport-related information is key to delivering good workshops.

Reflection 5: effective management of group sessions

Working with groups of young players brings with it a number of issues. Firstly, as mentioned above, the players often copied each other's responses. This attempt to avoid answering questions is an example of social loafing (Latane, Williams and Harkins, 1979), where certain members of the group are not motivated to contribute. Part of this might have been due to the players not understanding the benefits of engaging with the materials, which were not made as explicit as they could have been. Also, it seemed that some of the players were not willing to admit that they had had good experiences, which may be a sign that I had not made them feel comfortable with speaking in front of the group. This may have been overcome by allowing the players more opportunities to discuss issues in earlier sessions. However, my limited experience and confidence in delivering workshops meant that I followed a fairly rigid timeline that did not permit much time for such discussions.

Reflection 6: matching delivery materials to the appropriate age-group

The content of the workshops was sometimes too difficult for the young players. This was the consequence of two key factors. Firstly, I did not know the players before workshop delivery and was therefore unaware of how to 'pitch' the sessions. Secondly, I had little experience of delivering group workshops and was not aware of how much content I could fit into the one-hour sessions; this was despite being given a substantial amount of support from my SE supervisor. At the beginning of the process, my enthusiasm to tell the players what I knew and cover content in detail often resulted in the players not being able to understand the nature of the work or the interactive pen and paper tasks, and consequently they quickly became bored. Towards the end of workshop three it became apparent that my delivery methods were not fully effective. I became more flexible as the workshops progressed, allowing the content to be more fluid whilst gaining more input from the players. This flexibility enabled the players to have more time to engage with the content. This happened in consultation with my SE supervisor who deemed, after workshop three, that I was still trying to cover too much information for one-hour sessions. The consultations with my supervisor were invaluable in helping me feel confident and ready to deliver the workshops. Furthermore, Van Raalte and Andersen (2000) recognize the need for appropriate supervision and mentorship for practising applied sport psychologists. The session that had the most positive feedback was the one that contained some practical elements, which the players reported they enjoyed. Therefore, providing practical workshops may have been more appropriate for the groups of players involved.

Reflection 7: limited control over timing and room choice

The above issues were exacerbated by the timings of the workshops. The 'graveyard' shift after training was not conducive to good learning and often the players complained of being tired. In addition, the classroom set-up was inappropriate for the size of the groups resulting in rows of players sitting behind each other who could not always see the content being delivered. The room was also too warm for the players, which was not the most effective environment.

8.5 Needs assessment – season two

A season on, I had worked closely with the scholarship and academy squads through one-to-one consultancy with three players and attendance at several training sessions. Consequently, I now felt more like an integrated part of the sports science support team. Through regular attendance and talking to the coaches in the off-season, good working relationships were established with most of the coaches. I now had increased access to members of the coaching and support team, and was able to gain a more thorough and specific needs assessment. This process consisted of a meeting with the coaching staff regarding the previous season's performance profile, reviewing the performance evaluation sheets, match and training observations, and a discussion with the first team coach regarding what skills they needed in senior professional players.

Meeting with coaches and sports science support staff

This meeting addressed the season one performance profile and the staff updated the psychological qualities. During season one the 10 qualities used in the profiling had been agreed as mental toughness, footwork, grip and carry, passing, decision-making, playing the ball, speed, strength, tackling and communication. For season two, discussions resulted in changes to the 10 qualities being used for profiling to catch/carry/grip (one quality), tackling, aerobic fitness, upper body strength, lower body strength, commitment, T-CUP (thinking clearly under pressure), confidence, speed and posture. The coaches rated the most important qualities and many agreed that confidence was the key for all players between the ages of 13 and 16 years old, as the players often showed signs of low confidence when the teams were losing or when mistakes were being made. I spoke in detail to the new members of the sports science support staff to establish how the psychology sessions would fit in with their new training schedule. It was agreed that having the support staff at the psychology workshops sessions would be useful to the players, and that some of the workshops (for example the goal setting workshop) could be matched with the aims and objectives of the new periodized physical training schedule.

Discussion with the head coach of professional players

This brief discussion took place as part of a meeting with the coaching staff. The head coach suggested that at the senior level he needed players who were confident, self-aware and able to evaluate their own performance.

Observations

I was fortunate enough to be able to attend many training sessions throughout the previous season, and attended many matches for each of the four age groups. This proved invaluable for increasing knowledge of the sport, building rapport with the players and coaches, and understanding the needs of the players.

These sessions focused on observing the squad's overt behaviour and had the specific aim of investigating whether the coaches' interpretations (e.g. players' confidence decreases quite easily) matched the behaviours the players exhibited in training and competitive matches. To do this, I observed aspects of non-confident behaviours (e.g. reduced eye-contact and poor posture; Greenlees, Bradley, Holder and Thelwell, 2005a; Greenlees, Buscombe, Thelwell, Holder and Rimmer, 2005b) and noted their frequency and timing.

It became clear that many of the players exhibited poor body posture following parts of the match when they had not performed to their best, although this was only for short periods during the game. Greenlees *et al.* (2005b) argue that dropping the head reflects non-confident body language. I did not observe the 'non-confident' body language during the training sessions, suggesting that lapses in confidence were match-related rather than training-related.

8.6 Intervention and monitoring – season two

Development and delivery of workshop programme

The assessment showed that there were a number of issues that could be targeted such as confidence and thinking clearly under pressure. Hardy, Jones and Gould (1996) suggest that confidence could be one of the most powerful qualities for a performer and that it may be an important factor in the explanation of performance inconsistencies (Hardy, 1996). Therefore, the workshops aimed to provide tools to help to increase or maintain self-confidence. As the coaches expressed an interest in working on building confidence, I felt that this agreed goal would help the coaches to become more involved with the sessions in order to achieve the desired result with the players (Danish, Petitpas and Hale, 1993).

In addition to the coaches' input, season one's player evaluation forms showed clearly that they would like to do more 'practical' psychology work. I was disappointed with the slightly negative feedback from the players, suggesting that some of the classroom work was boring, and wanted to address the players' concerns in relation to the information gained for the needs assessment. I therefore decided to deliver more practically based workshops that would build on what had been previously delivered to them in the classroom. I arranged a meeting with my SE supervisor to present my ideas. This meeting focused on the logistics of delivering practical workshops rather than the detailed content of the workshops, which was the focus of the meetings in the previous season. I then presented a series of four workshops to the scholarship manager. Each of the workshops was discussed in detail, and the scholarship manager suggested ways in which the workshops could be incorporated into practical sessions. An overview of season two's workshops is presented in Table 8.2.

The workshops were all delivered in the training venue as part of the player's normal training sessions, earlier in the evening than the previous season's workshops. All

Table 8.2   Overview of season two workshops

Title	Age groups	Content	Activities employed
1. Performance profiling and goal setting	Twice(under 13s and 14s) and (under15s and 16s)	Filling in the pre-designed performance profile and setting two goals	45 minute interactive discussion led by sport psychologist, followed by interactive discussion led by physiologist (to address fitness goals). Final 45 minutes was spent in the gym identifying methods of achieving the fitness goals
2. Cue words for concentration	Twice(under 13s and 14s) and (under15s and 16s)	Developing new cue words and using them in a training environment	20 minute mini-lecture followed 70 minute practical session led by the respective head coaches
3. Imagery	Twice(under 13s and 14s) and (under15s and 16s)	Knowing what to image, when, how and why	20 minute mini-lecture followed by a 70 minute practical session acting out previous experiences whilst answering questions about their experiences
4. Acting confidently	Twice(under 13s and 14s) and (under15s and 16s)	How to react to pressure situations using specific behaviours	10 minute mini-lecture followed by 80 minutes acting out specific scenarios and role plays

squads, along with their respective head coaches, were expected to attend each of the four workshops. Each workshop was timetabled for one and a half hours to allow time for the practical content. There was approximately a two-week gap between each of the workshops to foster continuity.

The first workshop aimed to extend the first season's performance profiling workshop by getting players to rate their performance on the newly developed profile and by allowing the players to set goals as a result of their new profiles. The purpose of this was twofold. Firstly, Butler, Smith and Irwin (1993) identified that performance profiling is useful for enhancing self-awareness and goal setting, which in turn can increase confidence. Secondly, this aimed to raise awareness that, as confidence was identified on the performance profile, the coaches believed that confidence was a key psychological attribute of a successful performer. To further address the confidence aspect on the new performance profile, this workshop asked players to set two further goals based upon their fitness. The setting of the fitness goals was facilitated by the scholarship physiologist who was familiar with the training phase that the players were in. He also helped the players to be realistic in terms of what they wanted to achieve. The players were then taken into the gym as part of their training session and asked to identify the types of training they could do in order to achieve their new fitness goals.

The second workshop (cue words for concentration) related specifically to the 'thinking clearly under pressure' aspect of the newly developed performance profile. The workshop started with 20 minutes in the classroom exploring the use of cue words. The remaining 70 minutes were in the gym. The players were in groups of four. Players were asked to practise tackles using a relatively low-contact method as set out by the coach. The tackler was then asked to think of a cue word that they thought might help them before they tackled, as Butler (2000) suggests that cue words elicited by the athlete are the most effective. Each player shouted out a word, and then was asked to use it during the drill. The use of singular words supports Landin's (1994) encouragement to use brief, phonetically simple words when learning new skills. Players kept a log of which words they had tried, whether it had helped, and finally if so, how it had helped (i.e. did it aid concentration or help build aggression). This was then repeated with the ball handler trying different cue words. The use of cue words during practice is supported by Hardy, Gammage and Hall (2001).

The third workshop was based on imagery, as this technique has the potential to review previous performance accomplishments and can be used as a form of vicarious experience, thought to contribute to self-efficacy (Bandura, 1997). Paivio (1985) suggested that imaging successful performances and the rewards that follow could build an athlete's confidence. This workshop further developed the confidence and thinking clearly under pressure elements of the coaches' performance profile. This session started with 20 minutes in the classroom exploring the uses of imagery, including increased confidence and motivation. The remaining 70 minutes of the session took place in a gym and was delivered between squad coaches and myself. This workshop was based on Holmes and Collins's (2001) PETTLEP model and involved providing opportunities designed to encourage activation of the same neurophysiological processes as during physical practice in order to develop maximum functional equivalence (Decety and Jeannerod, 1996; Jeannerod, 1995). Players were asked several questions regarding their best ever tackle performance (including where it was, who it was against, what the weather was like and the state of the pitch). The players then acted out the tackle onto crash mats and were asked about how it felt, what they felt afterwards and what, if appropriate, they could hear or smell. This helped players develop vivid images which are said to aid performance (Ryan and Simons, 1982). The coaches were asked to encourage the use of imagery in the following training sessions.

The final workshop (acting confidently) aimed to examine how players react to pressure situations and then practise how they would respond to such situations. This workshop consisted of 10 minutes in the classroom discussing confident body language. The remaining 80 minutes was spent in the gym. For 5 minutes the athletes walked from the edge of the gym into the middle trying to reflect different types of behaviour (e.g. confident, tired, not confident and arrogant). The players then practised how they would like to look when walking out onto the pitch before a match. They then worked in pairs and were asked to hold each other down in tackles and act out how they would normally react and, secondly, how they would like to react. This drill aimed to help the players become aware of their reactions and subsequently the effect this can have on team-mates and the opposition.

8.7 Evaluation of intervention – season two

Evaluations from players

The players' evaluations were more positive in season two. The feedback suggested that they enjoyed the practical components, and that they were going to use the different techniques outside of the workshop sessions. For example, many identified specific times when they thought each of the psychological skills could be used within a match situation. Some suggested that they enjoyed getting to know their team-mates better during the workshops. Although this was not a primary aim, the coaches were very pleased with this response.

Reflections – season two

Reflection 1: integration of support staff

Workshop one (Performance profiling and goal setting) was attended by the squad physiologist. The integration of the physiologist was received very well by the players, who were able to ask questions and talk about the previous improvements they had made with the person they would normally approach about this matter. The increased participation by the players may be explained through the respect that they had built for the physiologist and myself. Respect is one of the building blocks for a good working relationship (Salacuse, 2000) and was also found by Jowett and Cockerill (2003) to underpin coach–athlete relationships.

The meeting with coaches, support staff and the scholarship manager prior to the workshops was also critical to the process. The scholarship manager talked about the potential value of sport psychology, and the other support staff appeared more enthusiastic about the workshop programme. Indeed, in the cue words session the first-team head coach led the session for 5 minutes, and the coaches and players seemed delighted. This helped the players engage with the content, knowing that what they were being taught also had application to the first-team players.

Reflection 2: including practical elements

The season two workshops differed from season one in part through the use of practical elements. The practical tasks were introduced in response to the players' evaluations and suggestions that they would like to be out of the classroom and my own reflections on season one. The workshops were also less structured, and although I planned key points within each workshop to be covered, the sessions were flexible enough to allow players' ideas to be aired and discussed.

Initially there were problems with engaging one of the coaches. This may have been caused by misconceptions that sport psychology was only for 'problem players', and that 'strong players' would not benefit from it (Pain and Harwood, 2004). However, after delivering the first practical workshop the coach seemed to change his attitude and began to ask me to do more work with their squad. I was very pleased with this reaction and later asked the coach why they did not seem to fully interact earlier. The coach reasoned that they did not see much value in the classroom work, a concept reinforced by Pain and Harwood (2004), who found that a lack of clarity concerning the services of a sport psychologist was a potential barrier. The coach later reported that he was pleased to see the players enjoy and learn from the practical workshop and suggested that he could see the value once they had seen the application to a match-related environment. The input of the coaches and support teams was invaluable to these workshops and the rapport with the staff led to a set of workshops that were both interactive and informative. At times players informally reported that they thought that they were in 'normal' coaching sessions rather than 'psychology' workshops. I was pleased with this feedback and believe that introducing the psychological skills within a physical task aided player understanding.

Additionally the new format ensured that everybody was involved, increasing potential opportunities for autonomy, and therefore enhanced intrinsic motivation (Deci and Ryan, 1985; Ryan and Deci, 2000). This may have helped create a mastery-based motivational climate that, as indicated in their evaluation forms, increased the players' intentions to practise their new skills in future training sessions and matches.

Reflection 3: trying skills in a controlled environment

In workshop two (Cue words for concentration) the training environment caused the players some problems. The players found it difficult to recreate the aggression experienced during an actual match. The players and I talked about the importance of introducing and developing new psychological (and physical) skills in a training environment. Despite the issues with training environments, the players agreed that practising in a slightly unrealistic environment was better than not practising at all. Abernethy (2008) suggested that, to enhance the development of physical expertise in young players, systems should be put in place that will maximize the frequency and likelihood of occurrence of the conditions that allow such development. This would appear to equally apply to psychological skills. Thus future training sessions needed to be arranged, with the support of the coaches involved, for this purpose.

During workshop four (Acting confidently), the players reported that the situations within the workshop were not realistic and therefore did not initiate the 'normal' feelings and thoughts that might hinder their ability to act in a specific manner. However, the squad talked about the ability to 'act' no matter what the situation and agreed that thoughts and feelings did not necessarily have to be reflected in their behaviours.

Reflection 4: being flexible during the workshop

During workshop four (Acting Confidently) some of the players also discussed their inability to control their behaviour when mistakes had been made. The importance of practising 'positive' reactions when mistakes are made in training was discussed. This activity led to discussion of other behaviours, such as shouting and self-talk, which I had not initially prepared for. Being able to adapt to the situation is an example of providing a good service, a factor identified by Anderson *et al.* (2004) as a key attribute of an effective sport psychology consultant. Some players talked about how they felt when others shouted at them, and by the end of the workshop they began to understand the relationship between the impact of verbal and non-verbal behaviours on others. This turned into a very important discussion for the players as they had not had a prior opportunity to discuss these important issues within team sports and effective communication has been shown to be directly related to group cohesion and team effectiveness (Yukelson, 1997).

Reflection 5: eliciting responses from players

Workshop three (Imagery) received the best written player feedback. The positive comments related to the practical elements of this workshop, and most said that imagery was a skill that they would now try and use as much as possible in training, with a view to eventually using it in a match situation. The workshops also focused on when each type of imagery would be most appropriate. Two players struggled with recalling their best performance, so the other players were asked to talk about these players' best performances. Subsequently, they were able to remember them with greater clarity. This flexibility in approach is something learned from the previous year's workshop where players would not admit to having good performances. Hearing others talk about their good performance seemed to provide the players with the confidence to talk about and imagine it.

Summary of reflections

To summarize, this two-season programme of educational workshops helped me to reflect on, and understand, a number of key issues for my applied practice. Specifically I learnt the importance of:

- being able to fit in with existing structures within the organization;

- building rapport with players and support staff;

- listening to others, with particular emphasis on responding to feedback from players and support staff;

- taking the time to design workshops based on the athletes' needs rather than exclusively from academic knowledge;

- producing materials that engage the performers and ensure that they have a positive experience of sport psychology, including using interactive tasks;

- developing the 'softer' consultancy skills, such as being enthusiastic about sport psychology, always being ready to help out and available for players to contact, and remaining committed to helping all the staff involved in the club.

8.8 Summary

The purpose of this chapter was to provide an insight into how an educational workshop programme aimed at rugby league junior age-group players was designed, implemented and evaluated across two consecutive seasons. In summary, in season one I delivered the workshop titles suggested by the scholarship manager. The workshops were predominantly classroom-based, consisting of mini-lectures followed by an interactive task. The feedback I gained from the players in season one was disappointing, but acted as a catalyst for me to restructure the programme for the following season. In season two, a more thorough needs assessment was conducted. This assessment led to a more effective and practical workshop programme, based on developing the players' knowledge of how to apply each of the psychological skills to training and game situations. Two key factors had changed between the two sets of workshops. The first was the involvement of the coaches and support staff. This led to a more coherent set of practical workshops, which complemented the specific training phases of the players. The players were better supported and, with the coaches' help, the sessions were integrated as part of their normal training regime in the training venue. The second key issue was my ability to respond to players' needs throughout the workshops. In season one the workshops were rigidly structured, which allowed very little room for players' ideas to be discussed. By season two I had developed more confidence to allow the players to contribute their ideas to the workshops and was able to facilitate discussion that I had not initially prepared for. These discussions led to better rapport between the players and myself.

Questions for students

1 How could you conduct a needs assessment with limited input from other support staff?

2 How might you gain access and build rapport with relevant support staff?

3 How can a sport psychologist gain sport-specific knowledge?

4 What are the key issues that need to be considered when working with groups of junior athletes?

5 When training to be a sport psychologist, discuss the attributes needed in an effective supervisor.

References

Abernethy, B. (2008) Introduction: developing expertise in sport–how research can inform practice. In: Farrow, D., Baker, J. and MacMahon, C. (eds), *Developing Sport Expertise: Researchers and Coaches put Theory into Practice*, pp. 1–14. Routledge, Abingdon.

Anderson, A.G., Knowles, Z. and Gilbourne, D. (2004) Reflective practice for sport psychologists: concepts, models, practical implications, and thoughts on dissemination. *The Sport Psychologist* **18**, 188–203.

Arksey, H. and Knight, P. (1999) *Interviewing for Social Scientists*. Sage, London.

Bandura, A. (1997) *Self-Efficacy: The Exercise of Control*. W.H. Freeman, New York.

Bull, S.J. (1995) Reflections on a 5-year consultancy program with the England women's cricket team. *The Sport Psychologist* **9**, 148–163.

Bull, S.J., Albinson, J.G. and Shambrook, C.J. (1996) *The Mental Game Plan: Getting Psyched for Sport*. Sports Dynamics, Eastbourne.

Butler, R.J. (2000) *Sports Psychology in Action*. Butterworth-Heinemann, Oxford.

Butler, R.J. and Hardy, L. (1992) The performance profile: theory and application. *The Sport Psychologist* **6**, 253–264.

Butler, R.J., Smith, M. and Irwin, I. (1993) The performance profile in practice. *Journal of Applied Sport Psychology* **5**, 48–63.

Danish, S.J., Petitpas, A.J. and Hale, B.D. (1993) Life developmental intervention for athletes: life skills through sports. *The Counseling Psychologist* **21**, 352–385.

Decety, J. and Jeannerod, M. (1996) Mentally simulated movements in virtual reality: does Fitts's law hold in motor imagery? *Behavioural Brain Research* **72**, 127–134.

Deci, E.L. and Ryan, R.M. (1985) *Intrinsic Motivation and Self-Determination in Human Behavior*. Plenum, New York.

Ghaye, A. and Ghaye, K. (1998) *Teaching and Learning Through Critical Reflective Practice*. David Fulton, London.

Greenlees, I.A., Bradley, A., Holder, T.P. *et al.* (2005a) The impact of opponents' non-verbal behaviour on the first impressions and outcome expectations of table-tennis players. *Psychology of Sport and Exercise* **6**, 103–115.

Greenlees, I., Buscombe, R., Thelwell, R., *et al.* (2005b) Impact of opponents' clothing and body language on impression formation and outcome expectations. *Journal of Sport and Exercise Psychology* **27**, 39–52.

Hardy, L. (1996) A test of catastrophe models of anxiety and sports performance against multidimensional anxiety theory models using the method of dynamic differences. *Anxiety, Stress and Coping: An International Journal* **9**, 69–86.

Hardy, J., Gammage, K. and Hall, C. (2001) A descriptive study of athlete self-talk. *The Sport Psychologist* **15**, 306–318.

Hardy, L., Jones, G. and Gould, D. (1996) *Understanding Psychological Preparation For Sport: Theory and Practice of Elite Performers*. John Wiley & Sons, Chichester.

Holder, T. (1997) A theoretical perspective of performance evaluation with a practical application. In: Butler, R. (ed.), *Sports Psychology in Performance*, pp. 68–86. Butterworth-Heinemann, Oxford.

Holmes, P.S. and Collins, D.J. (2001) The PETTLEP approach to motor imagery: a functional equivalence model for sport psychologists. *Journal of Applied Sport Psychology* **13**, 60–83.

Jeannerod, M. (1995) Mental imagery in the motor context. *Neuropsychologia* **33**, 1419–1432.

Jowett, S. and Cockerill, I.M. (2003) Olympic medallists' perspective of the athlete–coach relationship. *Psychology of Sport and Exercise* **4**, 313–331.

Landin, D. (1994) The role of verbal cues in skill learning. *Quest* **46**, 299–313.

Latane, B., Williams, K. and Harkins, S. (1979) Many hands make light the work: the causes and consequences of social loafing. *Journal of Personality and Social Psychology* **37**, 822–832.

Locke, E.A. (1996) Motivation through conscious goal setting. *Applied and Preventive Psychology* **5**, 117–124.

Locke, E.A. and Latham, G.P. (1984) *Goal Setting: A Motivational Technique That Works.* Prentice Hall, Englewood Cliffs, NJ.

Locke, E.A. and Latham, G.P. (1985) The application of goal setting to sports. *Journal of Sport Psychology* **7**, 205–222.

Maddux, J.E. (1993) Social cognitive models of health and exercise behaviour: an introduction and review of conceptual issues. *Journal of Applied Sport Psychology* **5**, 116–140.

Pain, M.A. and Harwood, C.G. (2004) Knowledge and perceptions of sport psychology within English soccer. *Journal of Sports Sciences* **22**, 813–826.

Paivio, A. (1985) Cognitive and motivational functions of imagery in human performance. *Canadian Journal of Applied Sport Sciences* **10**, 22S–28S.

Petitpas, A.J. (2000) Managing stress on and off the field: the littlefoot approach to learned resourcefulness. In: Anderson, M.B. (ed.), *Doing Sport Psychology*, pp. 33–43. Human Kinetics, Champaign, IL.

Poczwardowski, A., Sherman, C.P. and Henschen, K.P. (1998) A sport psychology service delivery heuristic: building on theory and practice. *The Sport Psychologist* **12**, 191–207.

Rugby Football League. (1994) *Know the Game: Rugby League.* A and C Black Ltd, London.

Ryan, R.M., and Deci, E.L. (2000) Self-determination theory and the facilitation of intrinsic motivation, social development, and well-being. *American Psychologist* **55**, 68–78.

Ryan, E.D. and Simons, J. (1982) Efficacy of mental imagery in enhancing mental rehearsal of motor skills. *Journal of Sport Psychology* **4**, 41–51.

Salacuse, J.W. (2000) *The Wise Advisor: What Every Professional Should Know About Consulting and Counselling.* Praeger, Westport, CT.

Scully, D. and Kremer, J. (1997) An educational approach: the design, implementation and evaluation of a psychological skills training programme. In: Butler, R.J. (ed.), *Sport Psychology in Performance*, pp. 147–176. Butterworth-Heinemann, Oxford.

Van Raalte, J.L. and Andersen, M.B. (2000) Supervision I: from models to doing. In: Andersen, M.B. (ed.), *Doing Sport Psychology*, pp. 153–166. Human Kinetics, Champaign, IL.

Weston, N. (2008) Performance profiling. In: Lane, A.M. (ed.), *Topics in Applied Psychology: Sport and Exercise Psychology*, pp. 91–107. Hodder Education, London.

Williams, S.J. and Kendall, L. (2007) Perceptions of elite coaches and sports scientists of the research needs for elite coaching practice. *Journal of Sports Sciences* 25, 1577–1586.

Yukelson, D. (1997) Principles of effective team building interventions in sport: a direct services approach at Penn State University. *Journal of Applied Sport Psychology* 9, 73–96.

9
Team Goal Setting in Professional Football

Richard Thelwell

University of Portsmouth, Portsmouth, UK

9.1 Introduction/background information

This case study provides an insight to some work that was conducted with a professional football club within the English football league. Prior to this particular season, the club had experienced little success both on and off the playing field for a number of years, and was in the second tier of the English league system. In addition to the lack of progress on the pitch, the club had suffered a significant turnover in managers in the preceding years, it had experienced a fluctuating financial situation where it had almost gone into administration, and it had crowd attendances that were much lower than many of the top clubs within their division. Put together, the club had experienced some dramatic organizational change (from boardroom all the way down to the playing staff) and it was a club that seemed (to me from an insider's view) to have lost its way and its vision of what it was trying to achieve. It was also as though the club knew that it wanted to be in the top flight of English football, but was unsure as to how it was going to be achieved. In essence, there seemed to be an *'outcome only'* approach with a lack of understanding of the *'processes'* that were required to succeed.

Prior to the season in question, I was also fortunate enough to have been involved in the previous season's campaign. The previous season had been very negative, where

Applied Sport Psychology Edited by Brian Hemmings and Tim Holder
© 2009 John Wiley & Sons, Ltd

players underperformed, the manager was changed towards the end of the season, supporters were disgruntled and the club narrowly avoided relegation to the third tier of the English league. Despite this, a number of important learning points were gained through being within a professional setting, communications with varying personnel (e.g. management, board, players, support staff) and through my own reflections, which were developed via a log book diary approach throughout the season. In short, the previous season had enabled me to develop a more detailed understanding of professional football, especially in terms of the structure, the culture and climate (use of the 'transfer' system, with players either joining or leaving), the organization and the development of relationships with playing and coaching staff. There were a number of other critical points, such as seeing how football clubs operate when experiencing difficulty. This is where the short-term outcome focus of the 'three points on a Saturday' statement became a real home truth for me (in English football three points are awarded for a win). Obviously I was aware of the importance of achieving success (as measured by the scoreline) in each match, but the first-hand experiences of being involved when three points were not achieved was essential to how my understanding of the game developed, and to how I went about my planning for the next pre-season.

Given the varied experiences of the previous season, the new season was being viewed by the club as the start of a new era. As such, throughout the off-season, a number of high-profile players with experience at the highest level signed for the club, and a number of changes at an organizational level took place that brought their own sets of issues (which are not addressed within this chapter – see Fletcher, Hanton and Mellalieu, 2006). With this in mind, it was critical that a positive environment was created within the club and that all personnel in the club were motivated appropriately to ensure that it was really going to be a new era. As a result, the purpose of the present chapter is to provide an insight into how I attempted to contribute to the 'new era' by ensuring that the squad as a whole were working towards agreed team goals throughout a competitive league season.

9.2 Initial needs assessment

Phase one: focus groups

The initial needs assessment employed differed somewhat from what I had become accustomed to with individual athletes, where a typical triangulation of interviews (with athletes and significant others where appropriate), performance profiling and observation enabled a grounding as to what the intervention would include. For a start this was a team sport and it was also a club that had received considerable investment to enter into a new era. As such, my first goal was to ensure that the psychology aspect of the pre-season was well managed and stimulating for the players. To do this, I needed to plan ahead and identify what my involvement was going to be. Given that coaches and players tend to be more receptive to the psychological aspects in the pre-season (Marchant, 2000) and that this is where the precedent for the remainder of the season can be set, it was essential that I had a framework for delivery in place. As a result, I used more of a top-down method (with the coaches) where a 'focus-group' approach (Bloom, Stevens and Wickwire, 2003) was adopted to identify what the coaches and manager perceived to be important for delivery.

The focus-group approach was something that I had only really read about in research methods texts and I had limited experience of conducting it in an applied setting. Despite this, I arranged a time for members of the coaching and support staff to come together to discuss what was to be included in the pre-season. Although a provisional timetable had been developed for much of the technical-based training, a total of six individuals (including myself) took part in the focus group, the key aim of which was to identify (a) what had been included in previous pre-seasons, (b) what worked well, and not so well, in previous pre-seasons and (c) what were we going to include in this pre-season (with an emphasis on the psychological and physiological/fitness testing aspects). From the outset of the session, it was made clear to all present that we all had our own ideas as to what we thought should be included and that that the pre-season programme would only work if each individual understood how the disciplines worked together. After approximately 90 minutes a consensus was reached relating to the number of psychology and fitness testing sessions that were going to be included in the pre-season programme. Interestingly, the conversation did not stop throughout the meeting and, to facilitate this, we employed a flip-chart to track all comments. Specifically, as each new comment was introduced, group members were given the opportunity to discuss the merits associated with it (e.g. whether it was appropriate or not, when it would have most impact). Throughout the focus-group meeting, we also managed to separate the pre-season period into a week-by-week schedule that enabled a more comprehensive overview of how we were going to build the players up to the start of the season. A total of six time slots were identified for 'squad psychology' sessions throughout the pre-season period, with the content being left to my discretion. Given my belief that all parties should contribute to the process and that myself, the coaches and players should all take responsibility (Ravizza, 1990), I suggested that the first session should be a 'squad profiling session' with the following five sessions addressing key aspects that resulted from the first session.

Phase two: squad profiling

In accordance with the recommendations forwarded within performance profiling research at an individual level (e.g. Butler and Hardy, 1992) and at a team level (e.g. Carron, Shapcott and Burke, 2007; Dale and Wrisberg, 1996; Weston, 2005), this session required the first-team squad (up to 18 players) and coaching staff to identify the characteristics of a team that would be able to secure promotion to the next league (English Premiership). This was agreed as an appropriate point of reference within the focus-group meeting conducted in phase 1 of the initial assessment and agreed by over 75% of the players prior to the commencement of the squad profiling session.

The squad and coaching staff were then split into four groups (mixing positions and experience within the club), in which they identified the qualities of the 'ideal squad' via brainstorming (cf. Butler and Hardy, 1992, p. 256). Following this, each group shared their list of qualities with the rest of the squad via a mini-presentation. Having concluded the four presentations, my role was to facilitate the development of the 'ideal squad profile'. This was conducted firstly by identifying qualities that appeared in all four presentations (seven qualities), then qualities that appeared in three presentations (five qualities) and then those that appeared in two presentations (four qualities). The squad agreed that any qualities that appeared on only one presentation were not worthy of inclusion. Having established a 'squad profile' that consisted of 16 qualities, the small groups then reconfigured to rate the squad as it currently stood. It was agreed that a score of 10 would resemble a squad worthy of promotion and a score of 0 would resemble a squad who were likely to experience relegation (see Table 9.1).

As a follow up to the squad profiling session, I briefly met with the squad prior to a 'fitness testing briefing session'. The purpose of this was to identify some key squad sessions that could use the identified time slots (see Initial Needs Assessment – Phase 1) and, more importantly, to address areas of the 'squad profile'. As a result, five sessions relating to 'Squad Issues – Controlling Controllables', 'Dealing with Distractions', 'Team Toughness', 'Developing Squad Confidence' and 'Developing Squad Goals' were agreed and delivered throughout the pre-season period. The purpose of the five sessions was three-fold. Firstly, each session addressed areas of the squad profile that the players had identified as an area requiring attention, thus enabling profile improvement (which was re-evaluated at an agreed later date – see Intervention and Monitoring). Secondly, by having sessions that reflected the needs of the players, they were more likely to experience greater focus towards the sessions and the squad as a whole (Weston, 2005). Finally, the sessions were designed in such a way that the players would be fully involved, and enabled a range of skills, such as communication, to be developed. This was important given the number of new players at the club.

Table 9.1 Performance profile developed during the pre-season phase

Quality	Meaning	Pre-season rating
Dealing with poor results	Showing an ability to deal and cope with a run of indifferent form – sticking together	7
Being able to keep winning	Showing an ability to be able to deal with the issues that relate to experiencing success	6
Showing we are unbeatable	Having a toughness as a squad that other people envy	5
Having confidence in each other	Knowing that each of us is a good player and that we believe in them	6
Showing pride in performance	Having a set of standards that we are happy with and that we want to maintain	6
Belief that we are physically strong	Knowing that when we go onto the pitch we are in top condition	7
Ability to know what we want	Having clear targets that we are working towards (both at a team and individual level)	5
Ability to communicate	Ability to communicate as a group and know how each player best communicates	6
Ability to help each other	Can identify when team-mates need support in training or a match	5
Having a good work ethic	Having a culture where we always want to develop – pushing each other	7
Positive response to going behind/conceding late	Being able to cope in a game when losing (especially late on in a game)	6
Ability to 'hold out'	Being able to close a game down when narrowly ahead	6
Generally being positive	Having a group that is always positive and able to overcome negativity within it	6
Having trust	Knowing that your team-mates are there for you	5
Ability to be flexible	Having an ability to cope with a change in formation during a game if players change	5
Playing with enjoyment	The top teams always seem to enjoy what they do	7

9.3 Interventions and monitoring

Despite the different approach to the 'initial needs assessment' to that which I had frequently employed, the main theme throughout the reported aspect of the intervention in this chapter relates to the development of team goals for the first phase of the competitive league season. In order to be able to identify relevant team goals, the final squad session, 'Developing Squad Goals' was critical. To facilitate the effectiveness of this session, I conducted some 'pre-session' work to ensure that the identified squad goals would be appropriate to what the club was trying to achieve as a whole. Here, a focus-group was re-employed in which the coaches identified their priority areas for the season. This was essential for a number of reasons. Firstly, I needed to ensure that the goals generated in the squad session were appropriate. Secondly, I was encouraging the coaches to develop responsibility for the goals and to reinforce them at all times. Finally, it was essential that the players saw that we (support staff, coaches and players) were operating together. The final point is particularly pertinent given the benefits of an integrative multidisciplinary approach to performance enhancement (Reid, Stewart and Thorne, 2004). Having understood the priorities of the coaching staff, I then met individually with some of the senior players, who had experienced promotion from the league in which they were in to the highest league. The reason for selecting these players was that they had experienced what we wanted the whole squad to experience (i.e. promotion) and as such they were more likely to be aware of the key goals that needed to be achieved to enable the overall goal to be attained. Ultimately, meeting with the coaching staff and some senior players gave me an insight into how I could manage the squad session and get a feel for the types and levels of goals that they perceived as relevant.

Although there was a need to be aware of the situation in which I was working (cf. Ravizza, 1990), I was also aware of the evidence base that supports the use of goal-setting within sporting environments. With this in mind, the use of goal-setting theory and research was important. With goals being defined as 'attaining a specific standard of proficiency on a task, usually within a specific time frame' (Locke, Shaw, Saari and Latham, 1981, p. 145) and team goals being defined as 'the future state of affairs desired by enough members of a group to work towards its achievement' (Johnson and Johnson, 1987, p. 132) the early phase of the intervention followed the principles forwarded by *Locke et al.*'s (1981) mechanistic theory. In accordance with this theory, goals are proposed to influence performance in four ways. Firstly, goals direct attention towards the important aspects of what is trying to be achieved. For example, by setting squad goals it was hoped that this would help members identify some of the key 'performance times' (e.g. last 5 minutes of each half) that needed work. A second benefit relates to how athletes mobilize their effort. Specifically, at a team level it would be anticipated that, from setting goals, greater effort would be exhibited within a practice environment. Thirdly, goals benefit persistence at a task. Given that the overall goal for the team was to achieve promotion, it was essential that the goals were appropriate to give persistence across the 8 month season. A final perceived benefit of setting goals is the manner in which they enable the development of new learning strategies. Thus within a team setting it may be that individuals learn from each other to enhance both individual and team performance. As reported previously, it was hoped that the approaches taken for the squad sessions would enable skills such as communication to be developed.

Despite the vast amount of research examining goal setting within individual settings, it was important to be aware that Locke and colleagues work did not make any predictions relating to the effectiveness of team goals when compared with individual goals (Locke and Latham, 1990). However it is proposed that goals set at a team level can be beneficial for the collective performance (Brawley, Carron and Widmeyer, 1992), cohesion (Brawley, Carron and Widmeyer, 1993; Holt and Sparkes, 2001; Schmidt, McGuire, Humphrey, Williams and Grawer, 2005) and collective efficacy (Greenlees, Graydon and Maynard, 2000) of a team. Further to this, despite the limited studies that have examined group goals and performance, the general consensus is that they are beneficial (e.g. Johnson, Ostrow, Pema and Etzel, 1997; Lee, 1989).

A second contrasting theoretical approach is the cognitive explanation forwarded by Burton (1983) that focuses on the benefits of setting goals for athletic performance. Specifically, Burton argued that goals are linked to the anxiety, confidence and motivation that athletes experience. In particular, it is not uncommon for athletes to focus on outcome-related goals (which also reflects what I thought the club did in previous years). Unfortunately, an emphasis on such goals can lead to unrealistic expectations that in turn can lead to decreased confidence, increased anxiety, reduced motivation and ultimately lower performance. Initially it was thought that performance-focused goals were more appropriate than outcome goals due to the greater control and flexibility that they provide the athlete. It was proposed that they enable more realistic expectations that enhance the development of confidence, motivation and performance. In contrast, Kingston and Hardy (1994, 1997) questioned the benefits of performance goals and suggested that they were based on the end result (albeit related to absolute or self-referenced performance levels). Instead, Kingston and Hardy suggested performance goals should be split into two categories. These include 'performance goals' (those focused on the attainment of performance standards) and 'process goals' (those focused on the development of strategy, form, and/or technique). With research support being developed for the use of the process goal approach to enhance cognitive states (e.g. Burton, Weinberg, Yukelson and Weigand, 1998; Zimmerman and Kitsantas, 1996), there is also evidence reporting the benefit of multiple goal setting strategies, where the use of each type of goal in combination, when appropriate, has been reported to enhance performance (e.g. Filby, Maynard and Graydon, 1999; Weinberg, Burke and Jackson, 1997).

As reported, the general theoretical underpinnings to goal-setting interventions have received widespread research attention (albeit largely from an individual perspective). With theory in mind, the intervention also followed guidelines provided by Widmeyer and DuCharme (1997) for the setting of effective team goals. The first guideline was with reference to setting long-term goals first. This was achieved at the very outset of the pre-season and was agreed as being 'automatic promotion to the higher league'. The second guideline related to setting clear short-term goals to enable the achievement of the long-term goals. It was agreed that there would be four main goal-setting periods throughout the season (August to October, November to December, January to February and March to May), with goal review and re-setting sessions scheduled for those periods. It is also important to note here that there were not too many goals set for any period. As with the goal setting intervention forwarded by Marchant (2000), there were never more than five goals set for each period, with each goal included being decided and agreed upon by the players. The third guideline related to all members being involved with the goal-setting process. Although this was largely adhered to, as stated previously, I met with some of the senior players to get an insight into what some of the key goals could be. The fourth guideline was to monitor progress towards the team goals. This was achieved in part by having four goal-setting periods and goals that were measurable where a tracking of progress towards the goal was possible. The penultimate guideline was to reward progress towards team goals. The coaches had some input with regard to this issue and they suggested that when goals were achieved the players would be able to choose what would be included in training on a low-intensity training day. The final guideline related to the development of team confidence from team goals. This was addressed by ensuring that the goals were at the appropriate level of difficulty to enable achievement and as such required input from the coaching staff where appropriate. To establish the appropriate level of goal difficulty, I used the information developed from my second focus-group with the coaches and individual meetings with senior players to ensure that the goals being set were appropriate. For example, one player commented that it would be totally worthless setting a goal to be in the top six for the season given that they did not want to be in the play-offs at the end of the season, but wanted straight promotion. Another player suggested that the goals should be set as though we were a Premiership team and that nothing other than the top two would do. Further to this, the senior players and coaches talked of the importance of achieving 'momentum', especially in the first few games of the season. Thus, it was achieving this positive 'momentum' that provided the level of difficulty associated for each goal (Jones and Harwood, 2008).

Based on the available theoretical, research and practitioner evidence on the potential benefits of setting of team-focused goals, the intervention presented in the following section outlines the process that I went through to be able to identify the key goals for the first goal-setting period (August to October). In keeping with the suggestions from the literature, the goals were driven by the notion of multiple goal-setting. Table 9.2 shows for each goal the goal type, the target for each goal, the strategies to achieve each goal and the methods of measurement (evidence). It is also important to note that all goals were set with the long-term goal of achieving automatic promotion in mind.

The first goal was evaluated by the coaching staff, who took charge of recording goal progress throughout the 10 game period (and the season as a whole) on a wall chart. The chart was placed in the canteen area at the training ground (in a position visible by every member of the playing, coaching and support staff) and enabled the plotting of points achieved for each game throughout the season. Further to this, the coaching staff plotted the points that they felt needed to be achieved to secure promotion at the end of the season, so that there was a running comparison between what was perceived to be needed and what was actually being achieved. Ultimately, goal 1 was achieved and the team were ahead of schedule by the first review date in terms of points achieved. Also, the team were unbeaten at home prior to the first review date. The second goal was fitness-based and was primarily identified and directed by the club physiologist. Feedback from the physiologist suggested the strategies to be effective for the development of both forms of fitness and the results indicated the goal to be achieved. This goal also included elements of flexibility given that only those players who conducted tests both pre-season and at the first re-testing date were included in the analysis. It is important to note that there were some players who joined the club towards the end of the pre-season period and as such did not benefit from the pre-season training. Such players were, however, included by the physiologist in further test re-test analyses of fitness.

The third goal was primarily identified by some of the senior players and coaching staff as being a key factor for successful teams. It was agreed at the outset of the season that setting goals and monitoring specific performance components at 'key moments' was important. My role here was to help the players (primarily in their playing units of defenders, midfielders and forwards) identify concentration and awareness strategies that enhance focus in the last 5 minutes of a half. The main strategy used (due to its relative ease) was that of a verbal statement triggered by the goalkeeper to enable heightened focus in the defenders for the final minutes of the half. From this, the midfielders were also made aware of the state of the game and finally one of the forwards would then take on further defensive duties. Put together, it was agreed with the coaches that, when a certain point in the game was reached (e.g. last 5 minutes of a half), where appropriate, the formation of the team may change to ensure that the scoreline was protected. This was especially the case when the team were leading by only one goal. In terms of goal achievement, the goal was achieved and the team conceded fewer goals in the final 5 minutes of each half in the first 10 league matches

Table 9.2 Overview of the squad goals set for the first phase of the season (August – October)

Goal	Goal type	Goal	Strategies	Evidence
1	Performance: short-term	To be in the top two division places	1. To identify and get the average number of points achieved by the teams getting automatic promotion for the previous five seasons	1. Develop wall chart showing where the team needs to be at the end of the season based on data from previous five seasons
			2. To identify and get the average number of points achieved by the teams achieving automatic promotion after 10 league games for the previous five seasons	2. Get statistics based on previous five seasons and plot progression line (for blocks of 10 league games) on the wall chart developed for strategy 1
			3. To remain unbeaten in home fixtures before the next review date	3. League records at the next review point
2	Performance: short-term/ fitness	For the first team squad to have improved their fitness results (both aerobic and anaerobic) by 10% for the next test	*Primarily identified by club physiologist:* 1. To adhere to squad training drills 2. To adhere to individual training programmes 3. To adhere to prehabilitation and rehabilitation programmes (training and match-day)	Aerobic and anaerobic fitness data as measured by club physiologist – feedback to the squad

3	Performance	To concede fewer goals than [team that won the league in the previous season] in the last 5 minutes of each half across the first 10 league games	1. To identify the number of goals conceded by [team that won the league in the previous season] in the last 5 minutes of each half across the first 10 league games 2. To develop strategies to enhance concentration and awareness when in the last 5 minutes of each half 3. To conduct training sessions directed towards developing performance strategies when in the last 5 minutes of each half	1. Statistics based on number of goals [team that won the league in the previous season] conceded in the last 5 minutes of each half across first 10 league games 2. Player (and coach) feedback on effectiveness of concentration and awareness strategies 3. Number of sessions specifically directed towards performance strategies and coach/player feedback on effectiveness of such sessions
4	Process	To be more assertive and tough as a team during matches	1. To identify some of the characteristics of an assertive and tough team 2. To identify strategies to enhance assertiveness and toughness on the pitch	1. To revisit the performance profile from pre-season and develop with a tough, assertive team in mind 2. To have developed, implemented and evaluated some strategies designed to enhance assertiveness and toughness

compared with the team that had won the league the previous season over the same time period. It would also be appropriate to note, however, that they also scored more goals throughout the period compared with the team that had won the league the previous season. One note of caution should also be made here with reference to the third strategy set for this goal. Goal setting literature (e.g. Burton, Naylor and Holliday, 2001) would suggest that the strategy should be more specific, for example, stipulating the number of training sessions to be conducted on the issue. However, following discussion with the coaching staff, it was decided that an exact number of sessions was inappropriate given the time spent training in the early part of the season (i.e. there is often a clustering of fixtures with league and league cup fixtures occurring) and that training also tends to follow the needs of the previous game and preparation for the next game. The coaches did acknowledge the importance of the strategy and agreed to implement the strategy where appropriate and within other sessions where possible.

The final goal was player-driven and set following one of the very early games in the season [similar to what Marchant (2000) referred to as a roving goal]. The team were on their way to experiencing their first defeat of the season to another team who were also labelled as 'promotion contenders'. Towards the final quarter of the game, the deficit was overturned and the team ended up winning the game. In the review of the performance, several senior players commented on the lack of assertiveness and toughness shown by the team as a whole in addition to individuals. As a result, during a training session in the following week, the squad explored the issue and it became the fourth goal of the first goal-setting period. In addition to reaffirming what some of the qualities held by an assertive and tough team could be (where there was also a revisit to the pre-season squad session on 'Team Toughness'), there were some on-field strategies such as the use of specific 'tough' statements and coming together during lengthy breaks of play that were developed to enable the players to refocus and help each other to adopt an assertive and tough approach (Miller, 1997). Primarily, the main drivers of the strategies were the 'spine' of the team [e.g. goalkeeper, central defender(s), central midfielder(s), forward(s)], who enabled communication between the smaller units of the team. Although this goal was harder to measure, the strategies were employed and the players did identify a number of common characteristics that could be associated with the assertive and tough teams. Further to this, it could be argued that the goal was achieved via alternative forms of measurement. From gaining some informal feedback from players, both individually and collectively, they perceived themselves to be tougher and more assertive (which may have been as a result of being top of the league). Also, through my observations and those of the coaching staff, there was a perceived development in the assertiveness of the team (which also may have been due to the increasing influence of the senior players who had joined the club and the positive results). Finally, and probably the best marker of the goal being achieved, were comments from opposing managers and players in the media that related to the resilience shown by the team. This suggested that, even when the team were not playing well, or were losing within the game, they were viewed as a team who would go until the 95th minute (we used the notion of playing to the 95th minute to ensure that focus and communication were maintained to the end – see goal 3).

In summary, the four goals identified were appropriate to the first phase of the season for two key reasons. Firstly, they were all developed with reference to theory and research and adhered largely to the SMARTER approach to setting goals (Locke *et al.*, 1981). Secondly, the goals were all underpinned by a number of reinforcing strategies (see Marchant, 2000) and identified and set via the input of players, coaches and myself.

9.4 Evaluation of intervention

As documented in recent literature (e.g. Anderson, Miles, Mahoney and Robinson, 2002), it is critical to evaluate the effectiveness of support work. Although evaluations were conducted throughout the whole season, this section will focus purely on the evaluation conducted after the first phase of the season. The evaluation was conducted via two approaches. Firstly, a series of meetings between two other support staff, one of the coaching staff, a selection of senior players and myself were held where consultant effectiveness (quality of support) was discussed. Although not evaluated directly in the manner suggested by Partington and Orlick (1987), the purpose of the meetings was to reflect on the psychological work (and other scientific work) by means of discussion. Here the key components of the discussions related to perceived impact of the goals (from the perspectives of myself, and the players), adherence to the goals and appropriateness of the goals. Given the perceived focus on the outcome of performances and the difficulty associated with changing this mindset, the staff were content and perceived the work to be effective so long as success was being achieved. As such, it may well have been that the work was not effective at all and it was mere coincidence that success was achieved at the same time as the psychological support was provided. One could suggest that this is a typical example of the outcomes achieved dominating the perceived success of what we do. A second method of evaluation was via re-profiling. In addition to providing consultant evaluation, the above-mentioned selection of individuals came together as a small group to re-profile the squad on the qualities identified in the profiling session. Although not directly related to the goals set in phase 1, the rating of the qualities gave an insight to how the squad were progressing towards a team capable of achieving promotion. As can be seen from Table 9.3, there were some notable improvements in some of the qualities such as 'being able to keep winning', 'ability to know what we want' and 'positive responses to going behind/conceding late' that the individuals attributed to the goals set at a team level.

Table 9.3 Comparison of performance profile scores between pre-season and end of phase 1

Quality	Pre-season rating	End of phase 1 rating
Dealing with poor results	6	6
Being able to keep winning	6	8
Showing we are unbeatable	5	6
Having confidence in each other	6	7
Showing pride in performance	6	7
Belief that we are physically strong	7	7
Ability to know what we want	5	8
Ability to communicate	6	7
Ability to help each other	5	7
Having a good work ethic	7	8
Positive response to going behind/ conceding late	6	8
Ability to 'hold out'	6	7
Generally being positive	6	7
Having trust	5	7
Ability to be flexible	5	7
Playing with enjoyment	7	8

With the above in mind, I also evaluated the work in an informal manner via individual meetings with players, where I used a social validation approach that is often reported in applied research designs (see Martin and Pear, 2003) to ask about their perceptions relating to the psychological support, and ultimately to whether they thought the team goals were of use. My reason for doing this was three-fold. Firstly, based on the knowledge that 'athletes hate paperwork' (Beckmann and Kellmann, 2003, p. 338), I wanted to avoid situations that would require them to complete such documents. Secondly, I was aware that some players may not give feedback in 'team' settings due to a lack of confidence in such situations. Thirdly, through having developed a good rapport with many of the players and from working with them individually as well as collectively, I perceived that I would be able to gain a better insight into my effectiveness from such an approach. From this, I would also be able to gain a more rounded approach from the 'foot soldiers' (i.e. those whom I was trying to direct the work towards). The individual evaluation drew some interesting responses with the vast majority perceiving the work to be effective. Such examples included, 'it's pretty good, we're doing pretty well and I'm getting my goals and the team are getting theirs', 'I have set goals before but they were set for me . . . we have all had an input here and we kind of own them . . . that's so much better, we have a responsibility' and 'I always thought the head stuff was a waste of time but it's not really, the goals are really useful and it's the first start to a season where I feel totally with it, the fitness is good, we all get on and we are top of the league'. Such evaluations were really beneficial for me because they reinforced that the material was being presented in a user-friendly way and that the players were able to link their individual goals to their team goals. This again reinforces that the suggestions from Dawson, Bray and Widmeyer (2002). Specifically they commented that there are a number of goals that are evident within teams that contribute to each other: team goals, team goals for individuals and individual goals.

9.5 Evaluation of consultant effectiveness/reflective practice

Throughout the season in question, I managed to adopt a range of self-evaluation and reflective approaches to ensure that my work was as effective as possible. These included attendance at appropriate continued professional development activities, discussions with other psychologists (both accredited and probationer) and continued study of the area. Further to this, the main vehicle was via detailed reflective practice, with specific use of the frameworks advocated by Gibbs (1988), Ghaye (2001) and Johns (1995). While the frameworks proposed by Ghaye and Gibbs were particularly fruitful with regard to reflecting on 'what has just happened' within a consultancy setting and how to develop an action plan for future events, Johns's approach was beneficial with regard to reviewing my practice across the whole season. As reported by many practitioners and researchers (e.g. Andersen, 2000; Anderson, Knowles and Gilbourne, 2004; Petitpas, Giges and Danish, 1999), we as practitioners have a responsibility to be aware of the effectiveness of the processes that we employ within our delivery, and of how our practice philosophy develops (Lindsay, Breckon, Thomas and Maynard, 2007). Acknowledging this, and from the use of continued reflection, I was able to become more aware of my personal attributes, how they can be best employed within my practice for increased effectiveness (Anderson, Miles, Robinson and Mahoney, 2004) and how such a knowledge may benefit individuals whom I supervise (Knowles, Gilbourne, Tomlinson and Anderson, 2007).

The following provides an insight into some of the key issues that I reflected on in my work within the squad. Given that the initial assessment, intervention and evaluation have focused on the first phase of the season, the first two reflections relate to that period. In addition to these and due to a number of significant developments within my consultancy work throughout the season, the final reflection relates to some thoughts I had relating to the work I conducted across the whole period.

To begin with, it is probably fair to suggest that practitioners (myself included) tend to work with an individual or a group for a short period of time, and then do not see them thereafter. It may have been fortunate due to the contract that I had with the club that the work continued successfully throughout the season. However I also felt that it reflected the relationship I developed with the squad and associated individuals (coaching staff, support staff) and the professional philosophy I adopted that enabled me to have enhanced awareness of the environment I was in (Poczwardowski, Sherman and Ravizza, 2004). More specifically, as can be seen from the first phase of the initial assessment, I made a conscious effort to include the key personnel in the decision making process. Ultimately, by having the coaching staff and senior players on board, I had identified the key stakeholders and enabled them to be involved in their development. This is certainly a process that I would employ again should I be involved in similar situations, with the main reason being that I was empowering the players, giving them some ownership and letting them take responsibility, rather than telling them what they were going to do (as reported in the player evaluations in the 'Evaluation of Intervention' section).

With regard to the team goal setting, reflections were critical at the start of the season to ensure that appropriate goals were set, and importantly that the goal difficulty, strategies and methods of evaluating were also appropriate. In particular, while it could be perceived as elitist to single senior players out, the discussions conducted with such people enabled a much clearer insight into the level of goal difficulty that was appropriate from the outset of the season. Further to this, I was aware that it would be inappropriate to set too many goals at a team level [hence the decision to follow the recommendations from Marchant (2000) and cap at five for any period]. Interestingly, the highest number of goals at any one stage was four, with the main reason for this being that the players also had their own goals that were identified in individual meetings. This obviously meant that I had to ensure that the individual goals related to the achievement of the team goals and vice-versa so that the commitment and motivation to achieve them was maintained.

My final reflection relates to the success that the team were experiencing throughout the season. There were a number of occasions where I found myself contemplating whether the players and staff would have been as receptive had the team not had a good start to the season or had the positive momentum to the season stopped, and also whether I would have adopted the same approaches. As the season progressed I became aware that they would have been. The key reason for this was the manner in which I had set the tone for the season in the pre-season phase. I had not thrown lots of material at the players – far from it in fact. Instead, I had let them contribute to the content. They owned the sessions and I merely acted as a facilitator who could direct appropriately (with the support of the coaching and support staff). Therefore, had the season followed a different path, I would always have had the squad's comments from the pre-season to revert to. As it was, the season did not follow a negative path, but it was a key point in reflection because at some time in the future I may find myself in a similar situation but with a different calibre of playing staff with different objectives. Then my reflections on the successes and the approaches that were used to enable it will be vital.

9.6 Summary

There are many variables that influence the achievement of successful team performance in sport. Two of the most notable include the ability of the whole team to be aware of what they are trying to achieve, and sharing the same team goals for the duration of the competitive period (Males, Kerr, Thatcher and Bellow, 2006). With this in mind, applied sport psychologists often spend a great deal of time assisting in the team goal setting process.

This chapter has described how an initial assessment was conducted with a professional football squad prior to team goals being developed for the first phase of the competitive league season. A squad profile was generated that enabled the development of four team goals for the first phase of the season. The intervention phase demonstrated the importance of appropriate team goals, and then described how the development of team goals took place. Within this, the measurement and recording issues associated with team goals were also considered. Reflections on the effectiveness of the work emphasized the importance of involving all relevant individuals in the goal setting process.

Questions for students

1 Identify some relevant team goals in other sports.

2 How would you approach a team goal setting session in a sport of your choice?

3 Discuss how you would ensure that the goals set within a team are appropriate.

4 How can a sport psychologist ensure adherence to team goals?

5 Discuss how you would approach working with a team to enable an effective needs analysis.

References

Andersen, M.B. (2000) Beginnings: intakes and the initiation of relationships. In: Anderson, M.B. (ed.), *Doing Sport Psychology*, pp.3–16. Human Kinetics, Champaign, IL.

Anderson, A.G., Knowles, Z. and Gilbourne, D. (2004) Reflective practice for sport psychologists: concepts, model, practical implications, and thoughts on dissemination. *The Sport Psychologist* **18**, 188–203.

Anderson, A.G., Miles, A., Mahoney, C. *et al.* (2002) Evaluating the effectiveness of applied sport psychology practice: making the case for a case study approach. *The Sport Psychologist* **16**, 432–453.

Anderson, A., Miles, A., Robinson, P. *et al.* (2004) Evaluating the athlete's perception of the sport psychologist's effectiveness: what should we be assessing? *Psychology of Sport and Exercise* **5**, 255–277.

Beckmann, J. and Kellmann, M. (2003) Procedures and principles of sport psychology assessment. *The Sport Psychologist* **17**, 338–350.

Bloom, G.A., Stevens, D.E. and Wickwire, T.L. (2003) Expert coaches' perceptions of team building. *Journal of Applied Sport Psychology* **15**, 129–143.

Brawley, L., Carron, A. and Widmeyer, W. (1992) The nature of group goals in team sports: a phenomenological analysis. *The Sport Psychologist* **6**, 323–333.

Brawley, L., Carron, A. and Widmeyer, W. (1993) The influence of the group and its cohesiveness on perception of group goal-related variables. *Journal of Sport and Exercise Psychology* **15**, 245–266.

Burton, D. (1983) Evaluation of goal setting training on selected cognitions and performance of collegiate swimmers. Unpublished doctoral dissertation. University of Illinois, Urbana, IL.

Burton, D., Naylor, S. and Holliday, B. (2001) Goal setting in sport: investigating the goal effectiveness paradigm. In: Singer, R., Hausenblas, H. and Janelle, C. (eds), *Handbook of Sport Psychology* (2nd edn), pp. 497–528. Wiley, New York.

Burton, D., Weinberg, R., Yukelson, D. it *et al.* (1998) The goal effectiveness paradox in sport: examining the goal practices of collegiate athletes. *The Sport Psychologist* **12**, 404–418.

Butler, R.J. and Hardy, L. (1992) The performance profile: theory and application. *The Sport Psychologist* **6**, 253–264.

Carron, A.V., Shapcott, K.M. and Burke, S.M. (2007) Group cohesion in sport and exercise. In: Beauchamp, M.R.and Eyes, M.A. (eds), *Group Dynamics in Exercise and Sport Psychology: Contemporary Themes*, pp. 117–140. Routledge, London.

Dale, G.A. and Wrisberg, C.A. (1996) The use of a performance profiling technique in a team setting: getting the athletes and coach on the 'same page'. *The Sport Psychologist* **10**, 261–277.

Dawson, K.A., Bray, S.R. and Widmeyer, W.N. (2002) Goal setting for intercollegiate sport teams and athletes. *Avante* **8**, 14–23.

Filby, W.C.D., Maynard, I.W. and Graydon, J.K. (1999) The effect of multiple goal strategies on performance outcomes in training and competition. *Journal of Applied Sport Psychology* **11**, 230–246.

Fletcher, D., Hanton, S. and Mellalieu, S.D. (2006) An organizational stress review: conceptual and theoretical issues in competitive sport. In: Hanton, S. and Mellalieu, S.D. (eds), *Literature Reviews in Sport Psychology*, pp. 321–373. Nova Science, New York.

Ghaye, T. (2001) Reflective practice. *Faster, Higher, Stronger* **10**, 9–12.

Gibbs, G. (1988) *Learning by Doing: A Guide to Teaching and Learning Methods*. Oxford Brooks University, Further Education Unit, Oxford.

Greenlees, I.A., Graydon, J. and Maynard, I. (2000) The impact of individual efficacy beliefs on group goal selection and group goal commitment. *Journal of Sports Sciences*, **18**, 451–459.

Holt, N.L. and Sparkes, A.C. (2001) An ethnographic study of cohesiveness in a college soccer team over a season. *The Sport Psychologist* **15**, 237–259.

Johns, C. (1995) The value of reflective practice for nursing. *Journal of Clinical Nursing* **4**, 23–30.

Johnson, D.W. and Johnson, F.P. (1987) *Joining Together: Group Therapy and Group Skills* (3rd edn). Prentice-Hall, Englewood Cliffs, NJ.

Johnson, S.R., Ostrow, A.C., Pema, F.M. *et al.* (1997) The effects of group versus individual goal setting on bowling performance. *The Sport Psychologist* **11**, 190–200.

Jones, M.I. and Harwood, C. (2008) Psychological momentum within competitive soccer: players' perspectives. *Journal of Applied Sport Psychology* **20**, 57–72.

Kingston, K.M. and Hardy, L. (1994) Factors affecting the salience of outcome, performance, and process goals in golf. In: Cochran, A. and Farrally, M. (eds), *Science and Golf*, Vol. 2. pp. 144–149. Chapman-Hall, London.

Kingston, K.M. and Hardy, L. (1997) Effects of different types of goals on the processes that support performance. *The Sport Psychologist* **11**, 277–293.

Knowles, Z., Gilbourne, D., Tomlinson, V. *et al.* (2007) Reflections on the application of reflective practice for supervision in applied sport psychology. *The Sport Psychologist* **21**, 109–122.

Lee, (1989) The relationship between goal setting, self-efficacy and female field hockey team performance. *International Journal of Sport Psychology* **20**, 147–161.

Lindsay, P., Breckon, J.D., Thomas, O. *et al.* (2007) In pursuit of congruence: a personal reflection on methods and philosophy in applied practice. *The Sport Psychologist* **21**, 335–352.

Locke, E.A. and Latham, G.P. (1990) *A Theory of Goal Setting and Task Performance*. Prentice-Hall, Englewood Cliffs, NJ.

Locke, E.A., Shaw, K.N., Saari, L.M. *et al.* (1981) Goal setting and task performance. *Psychological Bulletin* **90**, 125–152.

Males, J.R., Kerr, J.H., Thatcher, J. *et al.* (2006) Team process and players' psychological responses to failure in a national volleyball team. *The Sport Psychologist* **20**, 275–294.

Marchant, D.B. (2000) Targeting futures: goal setting for professional sports. In: Anderson, M.B. (ed.), *Doing Sport Psychology*, pp. 93–104. Human Kinetics, Champaign, IL.

Martin, G.L. and Pear, J.J. (2003) *Behavior Modification: What is it and How to do it* (7th edn). Prentice-Hall, Englewood Cliffs, NJ.

Miller, B.P. (1997) Developing team cohesion and empowering individuals. In: Butler, R.J. (ed.), *Sports Psychology in Performance*, pp. 105–128. Butterworth-Heinemann, Oxford.

Partington, J. and Orlick, T. (1987) The sport psychology consultant evaluation form. *The Sport Psychologist* **1**, 309–317.

Petitpas, A.J., Giges, B. and Danish, S.J. (1999) The sport-athlete relationship: implications for training. *The Sport Psychologist* **13**, 344–357.

Poczwardowski, A., Sherman, C.P. and Ravizza, K. (2004) Professional philosophy in the sport psychology service delivery: building on theory and practice. *The Sport Psychologist* **18**, 445–464.

Ravizza, K. (1990) Sportpsych consultation issues in professional baseball. *The Sport Psychologist* **4**, 330–340.

Reid, C., Stewart, E. and Thorne, G. (2004) Multidisciplinary sport science teams in elite sport: comprehensive servicing or conflict and confusion? *The Sport Psychologist* **18**, 204–217.

Schmidt, U., McGuire, R., Humphrey, S. *et al.* (2005) Team cohesion. In: Taylor, J. and Wilson, G.S. (eds), *Applying Sport Psychology: Four Perspectives*, pp. 171–184. Human Kinetics, Champaign, IL.

Weinberg, R.S., Burke, K. and Jackson, A. (1997) Coaches' and players' perceptions of goal setting in junior tennis: an exploratory investigation. *The Sport Psychologist* **11**, 426–439.

Weston, N. (2005) The impact of Butler and Hardy's (1992) performance profiling technique in sport. Unpublished doctoral dissertation, University of Southampton.

Widmeyer, W.N. and DuCharme, K. (1997) Team building through team goal setting. *Journal of Applied Sport Psychology* **9**, 61–72.

Zimmerman, B.J. and Kitsantas, A. (1996) Self-regulated learning of a motoric skill: the role of goal setting and self-monitoring. *Journal of Applied Sport Psychology* **8**, 60–75.

SECTION C

Working with Support Staff

10

Role Development and Delivery of Sport Psychology at the Paralympic Games

Jonathan Katz

Consultant Psychologist

10.1 Introduction/background information

Representing one's country at a Paralympic/Olympic Games is an ambition for many people involved in performance sport – athletes, coaches, support staff, sport medics and sport scientists. The reality is that a relatively small number of people actually realize this ambition. To illustrate this, Great Britain had six official psychologists supporting the 2004 Athens Paralympic and Olympic Games, with three located in Athens and the other three located in the pre-Games Holding Camps (see later for a description).

This chapter will describe work that I undertook between September/October 2002 and December 2004, with a particular focus on the development of my role as Head Quarters (HQ) psychologist to Paralympics GB. One of the distinguishing features of the Paralympic/Olympic environment is that it is a multi-sport event, where a number of different sports all compete simultaneously under a single Games organizing committee, where the athletes, coaches and support staff live in a purpose-built Athletes' Village and the range of different sports for Britain are entered into the

Applied Sport Psychology Edited by Brian Hemmings and Tim Holder
© 2009 John Wiley & Sons, Ltd

Paralympic and Olympic Games by the British Paralympic Association (BPA) and the British Olympic Association (BOA), respectively. The Athens 2004 Paralympic and Olympic Games was organized by the Athens Organizing Committee (ATHOC).

For the Athens 2004 Paralympic Games there were 16 discrete groups that could access the HQ psychology service comprising 15 sports plus the HQ core team. The BPA established a group to manage the 15 British sports totalling nearly 30 individuals in Athens. This group or team of individuals comprised the following roles: management, administration, logistics, sport medicine and sport science. This team is referred to as the HQ (or core) team as it offers centralized support that can be accessed by any member of the overall GB delegation/team. The HQ psychology role was one of the roles within this team.

The BPA had established a multi-disciplinary group to develop and provide strategic sport science and medicine support in elite disability sport. I became the BPA's consultant psychologist to this group in the autumn of 2002. The consultant psychology role was a blend of working at a national strategic level and the provision of sport psychology support. The support role included two broad, interrelated areas: provision of performance support direct to athletes, coaches and other support staff and offering organizational support to sports, facilitating improvement in management structure and communication systems.

The consultant psychology role with the BPA provided the broader context from which the specific Athens HQ psychology role evolved. Thus the HQ role evolved from an existing role with the BPA. The significance of this was that relationships with a number of the sports and partner agencies were already established prior to establishing the HQ psychology service.

Detailed attention is paid to the pre-Paralympic Games preparation. Attending a pre-Games Holding Camp is one method that has been developed to achieve quality pre-Games preparation. The pre-Games Holding Camp is usually held in the few weeks immediately prior to the start of the Games with the end of the Camp overlapping the start of the Games.

The purpose of a pre-Games Holding Camp is multi-factorial and is structured in a similar way to the Games environment in that it is usually multi-sport and there is the presence of an HQ, or core, team to simulate the Game's 'performance environment'. The camp provides an opportunity for people to acclimatize to the expected Games environment in terms of heat and humidity, as well as jet lag and travel fatigue. The sports' respective training programmes usually focus on 'tapering' their training so that athletes arrive at the Games well rested, focused and highly motivated to deliver optimum performances.

The pre-Games Holding Camp for the 2004 Athens Paralympic Games was held in Cyprus with several sports choosing to prepare independently and transfer directly into the Games Village. There were two sites in Cyprus, Nocosia and Paphos, and a camp for two sports based in the UK. The BPA ran several preparation camps in Cyprus prior to the final pre-Games Holding Camp. This afforded the opportunity to trial all aspects of the process, management, administration, training facilities, transport and accommodation, which also resulted in the formation and development of working relationships across the core management and support team and between the core team and the sports. This resulted in a number of important relationships being established and pre-existing relationships being strengthened. Key among these relationships were those between each sport's team manager and/or performance director and myself. It was out of these relationships that individual programmes of psychology support in the build up to and during the Games would be negotiated.

Security is a major factor at a Games. Individuals with accreditation have access to various venues within the Games infrastructure that include the Athletes' Village, sporting venues and the official transport. I was not an accredited member of the support staff team for Athens, along with several other colleagues, as the accredited places tend to be over-subscribed. I was fortunate in being able to access the Athletes' Village on a day pass and to access sporting venues on a spectator pass. The conditions of accreditation and security influenced the parameters within which individuals could access me for support, especially when HQ psychology support was requested or required in two or more venues on a given day.

Non-accredited support staff were accommodated in apartments outside the Athletes' Village in an Athens suburb. The apartment was relatively close to the main 'Olympic Park' where the main athletics, swimming, cycling and tennis stadiums were located. It was between 60 and 90 minutes (one-way) from the Village and other sporting venues. Thus, incorporating the impact of travel times became a constant feature in the process of all work planning.

This chapter will explore and reflect on the role of developing and delivering sport psychology support at the Paralympic Games, and in doing so, will provide useful lessons for all sport psychologists and not only those few who undertake this specific role. The chapter provides insights into delivering applied sport psychology support under pressure, for both the sport psychologist and the athlete. The Paralympic/Olympic environment provides a wide range of demands and pressures impacting on athletic performance and the quality of service delivery across management and sport science/medicine support staff. An implication of this for sport psychologists is to consider how to operate within this kind of multi-sport high-performance environment and the content of sport psychology support delivery. These insights can also be generalized to other international competitions where sport psychologists are in attendance. Further, sport psychologists can help prepare potential Paralympians and/or Olympians more effectively by having gained an appreciation of the unique demands associated with the Paralympic/Olympic environment.

10.2 Assessing the requirements/needs of the HQ role

Personal preparation

I understood, as part of my BPA consultant psychologist role, that I would most likely be providing support at the pre-Games Holding Camps only. As time progressed towards the end of 2003/beginning of 2004, the idea that psychology support might be included at the Paralympic Games itself began to surface. As a result, I initially focused on preparing for the pre-Games Holding Camp environment. I reviewed my personal preparation as it became clear that the HQ role could be extended to include the Games environment.

The Athens Paralympic Games was going to be my first experience of attending a major, multi-sport event. I had previous experience as a performer, coach and psychologist of major single sport championships, such as world championships. One of my central tasks was to offer psychology support to individuals and squads preparing for the Games. I identified that I needed to gain more information regarding the 'Paralympic/Olympic' environment and the impact of this on performance (e.g. Greenleaf, Gould & Dieffenbach, 2001), for athletes, coaches, managers, other support staff and for me, as a psychologist and as an individual. The scale of the Games and that it is multi-sport are two key features that characterize the Paralympic environment.

I appreciated that I needed to undertake professional development to develop my understanding and appreciation of the Paralympic/Olympic environment. This was achieved by what, in hindsight, can be considered a bespoke 'programme' of education and continual professional development. The 'programme' comprised the following strands:

- One-to-one mentoring – UK Sport was offering a practitioner mentoring programme to support the development of psychologists working within Home Country Sports Institutes and Sport National Governing Bodies to appreciate and achieve elite, world class preparation and performance understanding. For me this involved a series of one-to-one mentoring sessions from the summer of 2003 to the spring of 2005 with a highly experienced psychologist who had extensive knowledge of the Paralympic/Olympic environment.

- Discussion with previous HQ psychologists – the HQ psychology role is unlike other roles with the exception, possibly, of an HQ role at a Commonwealth Games. Consequently there are limited accounts of provision at Games in the literature (Males, 2006). I arranged meetings with two psychologists who had undertaken the HQ psychology role and, between them, had experience of both the Paralympic and Olympic Games environment. These meetings enabled detailed discussions around the personal and professional demands of the role, the content and process of an HQ psychology support service and appreciating the impact on communication systems of a multi-sport GB delegation managed by a 'core' HQ team.

- Leadership training – the BPA developed, in association with an identified organization, a bespoke leadership training package for HQ staff and sports team leaders/performance directors. The leadership training programme comprised a three-day residential training course repeated on two occasions. I attended both courses where I participated on the first course as a candidate and was involved in facilitating on the second to offer consistency of information across the courses.

- Olympic awareness training – the BOA developed and ran a bespoke seminar specifically for first-time Paralympic/Olympic support staff. The seminar, for me, was helpful in encouraging candidates to appreciate the demands of the Paralympic/Olympic environment from a 'personal cost' perspective. It emphasized the importance of recognizing one's own strengths and weaknesses, and of proactively considering developing personal and professional coping strategies.

- Media training and 'what ifs' – contingency planning was an important feature of the pre-Games preparation, which takes the form of 'what if' planning (Brooks, 2006; Butler, 1996). The BPA organized training workshops where members of the proposed HQ team and sports team leaders/performance directors systematically considered coping strategies to manage anticipated 'crises'. The group was sub-divided into smaller groups. Each sub-group was provided the task of managing a 'crises' throughout the course of the day and that additional information detailing the development of the situation was gradually 'drip-fed' into the sub-group as the day unfolded. Each sub-group member had an assigned role within the task. The object of the exercise was to continue with the day's normal business while simultaneously managing the emerging 'crisis'. Concurrent with this was the presence of the media staff who would seek you out for comment throughout the day. The net result was insight into the fact that daily duties need to be maintained; that unexpected situations can arise and impact on achieving the daily duties; and that all of this takes place in the public eye via the media.

- Athens team leader recce – a 'recce' is a visit to an identified location with a specific purpose of researching the logistics associated with that location. In this case, the location was Athens, where there was an opportunity to research the potential demands of the Athens Games environment. The BPA organized an Athens team leader recce for February 2004. I was aware, by this time, that it was likely that I would have a role in Athens as well as at the pre-Games Holding Camps and so I requested to go on the team leader recce. This was important for two main reasons. Firstly, I would gain first-hand experience of the geography of the venues, the Athlete Village and the logistics between the venues. Second, this insight would help to tailor pre-Games psychology support. I attended the recce and both goals were achieved as a result. The information gained during the recce confirmed that the various venues in Athens were located some distance apart, with the anticipated implication that journey times between venues would be a factor. Specifically it was expected that, because of longer travel times, it was not going to be possible to attend multiple sites on a given day. Thus, psychology support provision protocols in person and remotely by telephone were developed.

Professional preparation

Identifying and clarifying the organizational aspects of the HQ psychology service in terms of content and process was central. The BPA ran several multi-sport preparation camps in Cyprus prior to the pre-Games Holding Camp in 2003–2004. A major benefit was that it was possible to assess the psychological needs of those sports that attended. Discussions were held with the sports' team leaders/performance directors confirming their needs regarding sport psychology support. The outcomes of these discussions were reviewed and provided clarity on broad issues that could arise requiring support from the HQ psychology service:

- organizational systemic issues associated with effective communication systems, decision-making process and 'conflict' management;

- interpersonal relationship issues that are borne out of the unique demands of the pre-Games Holding Camp, and the Games environments;

- individual difficulties in maintaining a constructive focus within the pre-Games Holding Camp and the Games 'pressure-cooker' environment.

Table 10.1 provides a breakdown of specific issues that the HQ psychology service anticipated would require support provision across the pre-Games Holding Camps and Games environments.

The pre-Games Holding Camp and Games environments presented different challenges in planning the HQ psychology service. A key advantage for the pre-Games Holding Camp was that both sites in Cyprus were known, with most aspects of the proposed services having already been successfully delivered at previous preparation/acclimatization camps. Further, the environment was generally more under the BPA's control, in that there was full access to the hotel/accommodation, sporting venues and transport, thus the provision of psychology support could be

Table 10.1 Anticipated support provision at the pre-Games holding camps and Games environments

At the holding camp
Focus on fine-tuning 'the final preparation' by:

- Reinforcing existing coping techniques and strategies.
- Reinforcing self-belief and self-confidence.
- Reinforcing key preparation and performance tasks.

Holding camp – anticipated areas for support/intervention:

- Managing stress – relaxation; helpful distraction; mood management; balanced training – rest and recovery programme; preparation and performance routines.

- Mental skills review – focusing and re-focusing skills (centring); unhelpful distraction management; visualization and/or imagery skills.

- Self-belief/self-confidence – affirmations and positive self-statements; reminders of past success to support performance confidence.

At the games

- Managing Performance Expectations.

- Crisis support.

- Managing individual reactions in response to the environment, inter-personal and individual demands.

- Post-event support associated with under-achievement, injury, over-achievement, 'my last Games', etc.

- A balance between focused activity and rest and recovery (calm/quiet vs active).

At both the holding camp and the games

- Providing a supportive 'listening ear'.

- Being a 'sounding-board' for others to check out the appropriateness of a course of action prior to implementation.

- Managing unexpected domestic issues from home.

- Supporting 'pressure-resistant' relationships.

- Knowing when to back off and abstain from providing active support: 'If the best thing I can do to support you is nothing, then I'll happily do nothing!'

delivered optimally. This contrasted significantly with the Games Athletes' Village and competition venues in Athens, as there would be no opportunity for a 'dry run'. This required the psychology service to be established and delivered under 'competition conditions'; that is, the HQ psychology service infrastructure needed to be set up concurrent with the delivery of sport psychology support.

The preparation for the delivery of the HQ psychology service in the Cyprus pre-Games Holding Camps had been an ongoing process for approximately 18 months prior to the pre-Games Holding Camps. Thus, the majority of the focus in the months leading up to the Games was devoted to preparing the service for delivery in Athens. It was decided that I should base the HQ psychology service in the Village and then attend venues where sports had specifically requested HQ psychology support. It was anticipated that I would spend time each day in the Village and travel to venues for part of the day, returning to the Village prior to the day-pass terminating each evening (day passes were valid from 9 a.m. to 9 p.m.). This was the anticipated protocol that was used to negotiate HQ psychology support plans with sports prior to the Games commencing.

10.3 Interventions and monitoring

I will refer to the recipient(s) of psychology support at the Games as 'the athlete(s)'. The reader should bear in mind that the recipient could have been an athlete, coach or support staff member. Also information has been altered in the interest of maintaining anonymity and confidentiality within any case illustrations presented.

The HQ psychology service was made available to all members of the GB delegation/team. Sport psychology support was provided to individuals associated with nine of the 15 sports as well as members of the HQ core team (both pre-Games Holding Camp and Games staff), following referral to the HQ psychology service. Thus, individuals from 10 of the 16 groups accessed the psychology service, showing a wide range of individuals seeking support. See Table 10.2 for information related to referral and access to the HQ psychology service that was circulated to all the sports.

A psychology support 'contact'/consultation was defined as a planned or spontaneous interaction between myself and athlete or other individual seeking support. The duration of these interactions varied from 15 minutes to 4 hours per 'contact' with an average range of between 50 and 120 minutes. The total number of discrete 'contacts' recorded during the pre-Games Holding Camp and Games was 132. The 132 contacts included 165 discrete issues that were presented, which clustered around eight broad areas (Table 10.3).

This chapter will focus attention on exploring two of the eight areas: 'interpersonal relationships' and 'organizational issues'. The delivery of psychology support related to the other areas is discussed in other chapters of this book.

Table 10.2 The HQ psychology service: referral process

Individuals can:

- Self-refer

- Be referred by a member of the squad's management and/or sport science and medicine support staff.

- Be referred by a member of the HQ staff – generally via the medical team (doctor, nurse, physiotherapist) or the HQ Sport Science support staff.

- A brief 10–15 minute interview will be carried out to assess the individual's needs and then to formulate the necessary follow-up support that might be indicated.

- Subsequent more lengthy 1:1 sessions can then be arranged to fit into the individual's routine.

- If part of an existing programme of psychology support, then the appropriate session(s) can be arranged without the need for the initial interview:

Accessible to whom:
Psychology support will be available to:

- Athletes
- Coaches.
- Managers.
- Performance directors/team leaders.
- Supporters.
- Individual squad's sport science and medicine staff.
- HQ administration and performance staff.
- HQ sport science and medicine staff.

Interpersonal relationships

These issues referred to difficulties individuals had within the following interactions – between athletes; between athletes and coaches; between athletes and their squad's broader support staff and management; between coaches and support staff from the same squad; between members of a sport (athlete/coach/support staff) and members of the HQ core team; and between members of the HQ core team. Issues included

Table 10.3 The type and frequency of presenting issues

Type of issue	Frequency
Interpersonal relationships	35 (21%)
Performance preparation	29 (18%)
General support	27 (16%)
Organizational issues	21 (13%)
Mood management	18 (11%)
Performance expectations	15 (9%)
Personal issues	11 (7%)
Media issues	9 (5%)
Total	165 (100%)

relationship difficulties directly associated with the athlete's sporting performance to issues associated with living in the Village and general interpersonal strains associated with living in close proximity with others. Having an appreciation of the core conditions within the counselling process, such as warmth, empathy and unconditional positive regard, positively influenced how I was able to develop effective psychologist–athlete professional relationships (Casemore, 2006; Dryden, 2006).

Case illustration

An athlete sought support with an under-performance issue. It transpired that a fellow athlete within the same squad had been experiencing difficulties. These difficulties were being generalized from the sporting venue into the Village, which impacted on the relationships in the shared living apartment. The athlete seeking support described this situation as distracting them from focusing on managing their downtime effectively, which was having a negative effect on their ability to concentrate and focus on 'their job'. The support offered the athlete an opportunity to express their thoughts and feelings. This was achieved by adopting a predominantly client-centred approach (Casemore, 2006) encouraging the athlete to express their concerns via active listening skills (Jacobs, 1985), and introducing external distraction management techniques such as seeking environments within the Village where they could be entertained to take their mind off their concerns and relax (Hemmings, 2003). Once achieved, the support then became more directed towards encouraging the athlete to remind themselves of the pre-performance routines (Holder, 2003) they had successfully practised previously. This offered the athlete a clear and tangible focus that helped them feel 'back in control'. The athlete also expressed feeling concerned about his or her friend and we discussed how it was possible to be concerned for someone else without letting that concern negatively impact on your performance. The athlete subsequently reported being able to regain a constructive focus and was able to concentrate on their own performance preparation (Hardy, Jones & Gould, 1996).

Organizational issues

These issues revolved around the overcoming of communication difficulties by helping to establish clearer processes and protocols, providing assistance in decision-making and problem-solving in association with the interaction across different groups (i.e. the individual sport and the BPA, and the HQ Core Team and ATHOC). Although the Games environment provides challenges for athletes, it is for the support staff that the organizational issues have greatest impact. For an overview of the sport psychology literature on organizational stress see Fletcher, Hanton and Mellalieu (2006).

Case illustration

An implication of security issues and limited accreditations is that each sport tends to be under-staffed in comparison to single-sport championships; thus the same number of tasks need to be managed and completed by fewer individuals within each sport. This situation has significant implications for staff working within the Games environment. Specifically, each support staff member is required to undertake multiple job roles such as coaching roles, management tasks and athlete care (particularly with more disabled athletes and/or athletes with a visual impairment) in a pressurized environment with minimal or no opportunities for rest. Consequently support staff are faced with challenges associated with role conflict, role ambiguity and fatigue (Cooper & Dewe, 2004).

Role conflict arises when an individual support staff member is faced with meeting two or more demands simultaneously. Role ambiguity occurs when the individual support staff member is unclear on what is expected from them in a given task and/or how to manage the process within a specific situation. The cumulative effect of fatigue on the ability to concentrate and perform effectively was a constant challenge. A stress-management approach was adopted as an intervention approach (Jones & Hardy, 1990; Palmer & Strickland, 1995). The specific techniques involved assisting the athlete/staff member with effective appraisals and prioritizing of tasks whilst simultaneously reviewing current coping strategies. The athlete/staff member was encouraged to take 'personal time' daily, even if it was only 30 minutes, to undertake self-nurturing activities (Lazarus, 1999; Palmer & Dryden, 1995).

10.4 Evaluation of interventions/role

It is helpful to define intervention within the context of the HQ psychology role, in terms of the organizational process and content of delivery. The Athens Paralympic

Games HQ role encompassed both the pre-Games Holding Camp sites and the Games site in Athens.

At the pre-Games Holding Camp sites

The management of the Cyprus preparation camps included an explicit process of pre-camp planning sessions and post-camp debrief sessions. The HQ psychology role evolved throughout this process. Feedback was sought from sports in the camps regarding how beneficial they found all aspects of the camps from training venues, hotel suitability, transport to sport science and medicine. Feedback related to HQ psychology effectiveness was obtained as part of a camp-wide satisfaction questionnaire that was augmented by interviews held between team managers and the camp director. The outcome of the feedback indicated that individuals who accessed the HQ psychology service reported that the support provided assisted them in honing their final Games preparation as well as managing pressure and distractions effectively.

A referral process for the Camps was established whereby the HQ psychologist would be located in a consulting room at a specified time daily for a drop-in session. HQ psychology could also be accessed via a discrete sign-up sheet in the camp's office. As shown in Table 10.2, brief initial interviews of 15 minutes duration would be carried out followed by further sessions if needed. The proposed referral process proved to be unhelpful. Drop-in sessions for psychology did not generate a single session. The sign-up sheets were only slightly more successful as some athletes seeking support may have felt uncomfortable accessing the sheets in a public office space. The brief initial interviews, as a process, did not work. It was concluded that athletes would seek psychology support if the psychologist was visible and approachable rather than 'stuck in a consulting room' (see Partington & Orlick, 1987). Further, athletes tended to want to address their concerns when disclosed rather than arrange an alternative session time.

At the games

Athletes who sought HQ psychology support were routinely asked how beneficial the support was that they received. The majority of feedback was that the athletes were coping more effectively with the issues they had raised.

There were concerns raised by some squad staff members on their own behalf and on behalf of their athletes regarding accessing HQ psychology support. One issue was the fact that there was one HQ psychologist, which produced challenges associated with where I located myself (e.g. in the Village or at a sporting venue and, if so, which venue). A second challenge that impacted upon accessing HQ psychology support was not being fully accredited, as illustrated by the following quote from my reflective log/diary: 'Shortly before leaving the Village, X spoke with me about the difficulties with athlete A and also mentioned that athlete B was beginning to struggle. This discussion was held at 8.50 p.m. while walking to the Village exit' (as the day pass access expired at 9.00 p.m.). A third challenge was the impact of geographical logistics. The journey times between venues and the Village meant that the anticipated flexibility with travelling between multiple sites each day was impractical and this placed a priority on the use of telephone support alongside face-to-face support.

In summary, the HQ psychology service was used multiple times each day and was only limited by environmental and time constraints. The over-riding impression is that it offered a valuable support to members of Paralympics GB in cooperation with the other HQ support staff.

10.5 Evaluation of consultant effectiveness/reflective practice

Personal management

I arranged for professional support prior to attending the pre-Games Holding Camp sites and the Games. This support included arrangements with my mentor and also a colleague with whom I undertook monthly peer supervision (Andersen, 2005). There was an arrangement that I could contact either or both colleagues. I did use this resource on a couple of occasions and, interestingly, found that knowing that I had that support was, in itself, a great source of support (Edwards & Baglioini, 1993; Folkman, Lazarus, Gruen & DeLongis, 1986).

I believed that I prepared reasonably well in terms of gaining Paralympic/Olympic information, with the acquired knowledge being useful in gaining a sense of the scale of a major multi-sport Games. Attending the Athens recce provided me with an opportunity to contextualize some of this information. I did experience a degree of anxiety related to the Games part of the HQ psychology role. As much as the pre-Games Holding Camp sites environment was known and familiar, the Athens environment was foreign and unfamiliar. It was a combination of the unfamiliarity and uncertainty of the Games environment, and the challenges of meeting the related high-performance demands that caused my concerns (Katz, 2006).

Reflective practice is encouraged in applied sport psychology (Cecil & Anderson, 2006). I kept a daily reflective diary (or log) related to the role of HQ pychologist throughout my attendance at the pre-Games Holding Camps and Games. This was maintained and achieved for a 31-day period that included attendance at the pre-Games Holding Camp sites in the UK, and Nicosia and Paphos in Cyprus as well as at the Games in Athens. I intentionally made the final entry to coincide with my first full day back home to provide the opportunity for final reflective insights. I established a clear protocol of completing the log daily; each entry was written 'exactly as I felt' at the time without re-writing and/or grammatical correction. In short, it was my intention to create an authentic record. The final log provided invaluable information that has been central in my being able to review the content and process of the HQ psychology service.

'Controlling the controllables' is a well accepted and commonly adopted working approach in sport psychology (Jennings, 1993). My early experience in Athens highlighted how many factors that impacted on me and my role as HQ psychologist were outside of my influence. I have acknowledged that I became negatively distracted by some of these factors, most notably geographic/logistical challenges and access availability due to limited accreditation. A direct result of keeping a reflective log was that I became aware of this process and consciously re-directed my focus onto more productive areas, which resulted in me generating a timetable that I shared with colleagues, informing them of my priorities and when I would be Village-based or sports venue-based. This schedule, although adapted to meet changing demands, promoted the best use of my time. Below is a quote from my reflective log on day 16 that illustrates this:

> Reflection: a better day personally for me today. Worked hard on focusing on the positives and generally working within the parameters I have . . . I still need to work harder on remaining disciplined and focused on tasks reducing getting distracted – particularly when multiple requests come in simultaneously.

The quality of interpersonal interactions and the impact of this on relationships affect the process of psychology support delivery (Poczwardowski, Sherman & Henschen, 1998). Thus awareness of transference and counter-transference dynamics is important for the psychologist (Timson, 2006). Transference refers to the situation when an athlete attributes his or her own thoughts and feeling to the psychologist. Counter-transference is when the psychologist attributes his or her own thoughts and feeling to the athlete (Feltham & Dryden, 1993).

In terms of counter-transference, I have mentioned that issues associated with access, accreditation and geography had a daily impact on delivering the HQ psychology role. This provided me with a daily challenge to ensure that I managed my own thoughts and feeling so that they did not negatively impact on my interactions with athletes and colleagues. Below is an entry from my reflective log describing the challenge I had in managing these issues:

> Reflection: I've just written in the first bullet point 'people on Day Pass accreditation'. It occurred to me that I am accredited and on official ATHOC lists, just not to the level of access that I'd prefer!!! This is a useful cognitive restructuring with, if I'm honest, a reality check. Perhaps if I'd reminded myself of this earlier, it might have reduced some of the frustration that I've felt when my work goals have been blocked by ATHOC officialdom.

The issue of not being accredited and gaining access to some athletes seeking support became intertwined in my thinking. Consequently there was a risk that I would become distracted by what I could not do rather than focusing on the areas, issues and athletes that I could make a constructive difference with. Once I recognized this, in the quote above, I was able to amend my work plan and practice accordingly.

The Games environment, in particular, is highly pressurized and intense. Consequently athletes' emotional experiences tended to be extreme: either elation or despair (i.e. they were emotionally labile). This emotional climate promoted intense undercurrents that affected the quality of interpersonal relationships (transference) with me. My experience and background in counselling and clinical psychology provided me with useful strategies. The first step was to be aware of this and to take the time to allow athletes to express their feelings before moving on to discuss coping strategies. Often, the process of 'emotional off-loading' was sufficient for athletes to move forward, as once they had got their feelings off their chest, they were aware of what they needed to do. Thus, building rapport through gaining empathic understanding was helpful. I found that taking the time to listen and understand, providing a 'calm and contained emotional space', was helpful in encouraging athletes to disclose their concerns.

Managing the organization and logistics

The decision for me to go to Athens as well as the pre-Games Holding Camps significantly influenced the preparation process. The HQ psychology role in Athens was a non-accredited position. This resulted in me living out of the Village in an apartment in an Athens suburb. Access to athletes and other clients was available on a day-pass basis in the Village from 9 a.m. to 9 p.m. Venue access was achieved through spectator tickets, which meant that I was not able to access the athlete areas. The official Paralympic sites (venues and Village) were spread out quite widely across Athens and the surrounding areas. The combination of living out of the Village, having limited athlete access and significant geographical distances between different venues produced logistical challenges.

The challenge of delivering HQ psychology support differed between the pre-Games Holding Camp sites and the Games. The organizational structures and systems in the pre-Games Holding Camps had been fully tested at preparation camps. This meant that the primary task for the pre-Games Holding Camp sites was to concentrate on delivering what had already been prepared. The implication for the Holding Camp HQ psychology service was that the organizational and logistical issues had already been resolved. This was, for me, reassuring, as I was familiar with the Holding Camp environment and associated demands, had established effective working relationships with colleagues and had tangible experience of delivering an HQ psychology service in that environment at the preparation camps. Thus, preparation could focus on ensuring optimum psychology support delivery with particular attention paid to managing potential unexpected issues alongside the planned routine psychology support programme. Consequently I felt somewhat more confident in preparing for the Holding Camp in comparison to the Games, where there were more unknown variables.

The delivery of HQ psychology support at the Games was more challenging. The combined lessons from the Cyprus preparation camps were included in the Games preparation. Given that it was not possible to replicate the Games environment with a view to testing out structures and processes, the specific demands and pressures associated with setting up and delivering the HQ psychology role were higher. Psychology support was provided in both planned and spontaneous consultations at sporting venues and in the Village. These sessions took the form of both direct and indirect delivery systems including one-to-one, one-to-two and group sessions. Remote support was also provided via the use of 24 hour telephone access, mobile phone text messaging and occasional e-mail. Telephone access was a useful mode of support as it provided indirect access when face-to-face contact was not possible. It was most helpful with those athletes who had already had face-to-face contact and an established professional relationship, and of lesser benefit without this relationship. Continuously reviewing and amending the HQ psychology delivery structures and process was an ongoing feature throughout the Games in response to changing demands and priorities.

This is illustrated by the following. I became increasingly aware of the need to impose a structure on my role by identifying priority needs both in terms of which sports/athletes to support and in which venues to do so. The following extracts from my reflective log (day 24) describe this further. Note, this was being undertaken on day 24 of 30, indicating how I was continually reviewing and adapting my working role:

> Arrived at the Village and went to find X at their request. X said it would have to wait due to another demand . . . I provided a suggested schedule of my movements for the final few days of the games. This schedule prompted an invitation to join a planning meeting with X, Y and Z. It felt good to have such a meeting.

These reflections led to a greater clarity of my role that, in turn, was more effectively communicated, which resulted in an improved focus on HQ psychology support delivery.

10.6 Epilogue

This chapter was completed following my return from the 2008 Beijing Paralympic Games, which provides an opportunity to offer further reflections on the role of HQ psychologist. Accreditation and experience of attending a previous Games are two notable differences between my HQ psychology role in Athens compared with Beijing.

Being accredited in Beijing meant that I was living in the Athletes' Village, had access to all of the sports venues and could make use of the official transport. In contrast to Athens, this enabled me to plan and deliver the HQ psychology service where I could access athletes and sports staff. Further, living in the Athletes' Village meant that I could provide psychology support, if needed, 'out of hours'. In contrast to Athens, being accredited resulted in my being able to manage the logistics of the HQ role with greater ease. As a consequence, I was able to focus my personal and professional resources on the HQ psychology role with minimal distraction from logistical issues.

In reviewing my reflections from Athens, I noted that there was a discrepancy in the level of my concern related to the pre-Athens Holding Camp and the Athens Games, where I felt somewhat more concerned about my role in Athens. I attributed this in Athens to it being my first Paralympic Games and my having no previous first-hand experience of that environment, thereby perceiving there to be many more uncontrollable than controllable factors. In Beijing, by contrast, I found myself feeling comfortable in the knowledge of having an understanding of both the range of issues and demands of the Paralympic/Olympic environment. Thus I was able to separate out the 'occasion' from the 'event', and maintain my focus constructively on the tasks associated with my role as HQ psychologist.

10.7 Summary

The HQ psychology role for the Athens 2004 Paralympic Games incorporated the provision of psychology support at the pre-Games Holding Camp sites and the Games itself. The role was multi-faceted with planning, preparation and organizational demands prior to the Games and a maintenance role related to these tasks during the pre-Games Holding Camp and the Games. It also emphasized the need to offer psychology support to any member of the GB delegation/team. The range of issues supported was wide, from organizational issues to supporting individual athletes' performances. Finally, the HQ psychology role was influenced by a series of wider organizational factors associated with accreditation, access to athletes and geographical logistics.

Questions for students

1 The Paralympic/Olympic environment provides a specific range of challenges for the psychologist. What are the implications for psychology support when preparing elite performers?

2 What are the ways a psychologist can prepare him or herself for a support role at a major championship?

3 What are the issues associated with referrals and athletes gaining entry to a psychology support service?

4 What are the challenges for the psychologist in meeting the wide range of athlete issues such as those described in this chapter?

5 What are the advantages of having a counselling/clinical psychology background when operating as a sport psychologist?

References

Andersen, M. (2005) *Sport Psychology in Practice*. Human Kinetics, Champaign, IL.

Brooks, J. (2006) Reflections on the Athens Olympics and Paralympics: my work as a sport psychologist working with equestrian. *Sport and Exercise Psychology Review* **2**, 35-40.

Butler, R. (1996) *Sport Psychology in Action*. Routledge, London.

Casemore, R. (2006) *Person-Centred Counselling in a Nutshell*. Sage Publications, London.

Cecil, S. and Anderson, A.G (2006) Reflections on the reflections. *Sport and Exercise Psychology Review* **2**, 46-48.

Cooper, C.L. and Dewe, P. (2004) *Stress: a Brief History*. Blackwell Publishing, Oxford.

Dryden, W. (2006) *Counselling in a Nutshell*. Sage Publications, London.

Edwards, J.R. and Baglioni, A.J. (1993) The measurement of coping with stress: construct validity of the ways of coping checklist and the cybernetic coping scale. *Work and Stress* **7**, 17-31.

Feltham, C. and Dryden, W. (1993) *Dictionary of Counselling*. Whurr Publishers, London.

Fletcher, D., Hanton, S., and Mellalieu, S.D. (2006) An organizational stress review: conceptual and theoretical issues in competitive sport. In: Hanton, S. and Mellalieu, S.D. (eds), *Literature Reviews in Sport Psychology*, pp. 321-373. Nova Science, New York.

Folkman, S., Lazarus, R.S., Gruen, R.J. *et al.* (1986) Appraisal, coping, health status, and psychological symptoms. *Journal of Personality and SocialPsychology* **50**, 571-579.

Greenleaf, C., Gould, D. and Dieffenbach, K. (2001) Factors influencing Olympic performance: interviews with Atlanta and Nagano Olympians. *Journal of Applied Sport Psychology* **13**, 154-184.

Hardy, L., Jones, G. and Gould, D. (1996) *Understanding Psychological Preparation for Sport: Theory and Practice of Elite Performers*. Wiley, Chichester.

Hemmings, B. (2003) Dealing with distractions. In: Greenlees, I. and Moran, A. (eds), *Concentration Skills Training in Sport*, pp. 43-54. The British Psychological Society, Leicester.

Holder, T. (2003) Concentration training for closed skills – pre-performance routine. In: Greenlees, I. and Moran, A. (eds), *Concentration Skills Training in Sport*, pp. 67-75. The British Psychological Society, Leicester.

Jacobs, M. (1985) *Swift to Hear: Facilitating Skills in Listening and Responding*. SPCK, London.

Jennings, K.E. (1993) *Mind in Sport: Directing Energy Flow into Success*. Juta & Co, Kenwyn.

Jones, G. and Hardy, L. (1990) *Stress and Performance in Sport*. Wiley, Chichester.

Katz, J. (2006) Reflections on the Paralympic HQ psychology service: Athens 2004. *Sport and Exercise Psychology* **2**, 25-28.

Lazarus, R.S. (1999) *Stress and Emotion: A New Synthesis*. Free Association Books, London.

Males, J. (2006) Reflections on Athens: delivering sport psychology provision at the BOA headquarters. *Sport and Exercise Psychology Review* **2**, 12-16.

Palmer, S. and Dryden, W. (1995) *Counselling for Stress Problems*. Sage Publications, London.

Palmer, S. and Strickland, L. (1995) *Stress Management: A Quick Guide*. Daniels Publishing, Cambridge.

Partington, J. and Orlick, T. (1987) The sport psychology consultant evaluation form. *The Sport Psychologist* **1**, 309-317.

Poczwardowski, A., Sherman, C. and Henschen, K. (1998) A sport psychology services delivery heuristic: building on theory and practice. *The Sport Psychologist* **12**, 191-207.

Timson, S. (2006) Reflections on Athens: delivering sport psychology at the BOA Cyprus holding camp. *Sport and Exercise Psychology Review* **2**, 20-24.

11

An Integrated Multi-disciplinary Support Service for an Injured Rubgy Union Lock

Sarah Cecil, Raphael Brandon and James Moore

English Institute of Sport, London, UK

11.1 Introduction/background information

The importance of sport psychology in the injury rehabilitation process is increasingly being recognized. Current understanding suggests that critical areas in this process that need to be addressed include effective rapport building, managing perceptions of rehabilitation and maintaining motivation through effective goal setting (Petitpas, 2002). Current research also assigns a key role for cognitions in the injury process (Brewer, 1994, Wiese-Bjornstal, Smith, Shaffer & Morrey 1998; Udry & Andersen, 2002).

Understanding the role that the sport psychologist plays in effective multi-disciplinary teams is vital in the current sporting environment. In this chapter I will document my experiences as part of an effective multi-disciplinary team in the rehabilitation of an injured international rugby union player. The need to provide background on the organization, the practitioners, and the athlete involved, stems from the increasing call in the literature to understand more of the 'how' of sport psychology delivery (Andersen, 2005).

Applied Sport Psychology   Edited by Brian Hemmings and Tim Holder
© 2009 John Wiley & Sons, Ltd

The English Institute of Sport (EIS) was established in 2002 and provides a nationwide network of world class support services including sport science and sport medicine. The three EIS practitioners involved in this intervention provided physiotherapy, strength and conditioning, and psychology support. All three practitioners were employed full time for the English Institute of Sport and had rugby-specific experience ranging from six months to five years at the time of the intervention. This intervention was part of comprehensive support provided to the Rugby Football Union Women (RFUW) in the lead-up to the World Cup in 2006.

TJ – the athlete

TJ, a 36-year-old female lock, sustained injury to her medial collateral ligament (MCL) 11 months before the World Cup Final in 2006 from a collision during a breakdown in open play against Samoa in October 2005. The injury was diagnosed as two grade three strains to her MCL in her left knee.

In 1998, TJ had made the decision to base herself full time in England. Whilst pursuing her career as a banker in the City, TJ had played club rugby for Richmond. She gained her first cap for England A in the 1999/2000 season. During the World Cup Season of 2002 TJ suffered a knee injury and had a meniscectomy on her right knee. She felt on reflection that she had rushed her return to rugby during the following Six Nations tournament and this had impacted her ability to perform to her potential during that year's World Cup. She also felt that this had resulted in her having to watch rather than play in the World Cup Final in 2002, a situation she did not want to repeat four years later in 2006. When TJ presented with her injury in November 2005, her previous injuries on her right knee on impacted the rehabilitation of her new injury.

TJ had been working with the strength and conditioning coach for three years since 2003. The physiotherapist and I first came into contact with TJ in September 2005.

11.2 Initial needs assessment

At TJ's first session following her injury she presented a variety of concerns. Utilizing a cognitive–behavioural therapy (CBT) approach, socratic questioning was utilized to help clarify what thoughts were significant in TJ's mind. Socratic questioning, often viewed as the cornerstone of CBT, involves asking questions that the client can answer, which draws the client's attention to relevant information that might not be in the client's focus (Padesky, 1993). A number of negative thoughts were evident for TJ, including:

• Will I favour my knee?

• Will it be 2002 again?

- Should I still be playing rugby?

- Did I come back too early in the Six Nations in 2002, which impacted on my World Cup in 2002?

Identifying common thinking biases and faulty thinking patterns is part of the process of initial assessment/case conceptualization in CBT (Kennerley, 1997). With TJ there was evidence of common thinking biases, namely *catastrophization* and *jumping to conclusions* with the thought 'will it be 2002 again?' and *overgeneralization* with the thought 'should I still be playing rugby?' In addition, there was reference to beliefs about the impact of past injury management on her ability to be at her optimum playing ability during the 2002 World Cup.

A key part of the case conceptualization and treatment process in CBT is sharing the process and educating the client (Persons, 1989). The different types of common thinking biases that TJ displayed were pointed out to her and the 'hot thought' of 'Will I favour my knee?' was elicited. The term 'hot thought' in CBT is used to identify the automatic thought that is most connected to the mood that the individual experiences (Padesky & Greenberger, 1995). The hot thought can be viewed as the thought which is most emotionally charged.

The next step in case conceptualization is to utilize the five aspects of your life model (Padesky & Greenberger, 1995) to illustrate how the hot thought influences emotions, behaviours and physical reactions. The model suggests that presenting problems are studied by investigating thoughts, moods, behaviours, biology and environmental factors. Padesky and Mooney (1990) have demonstrated how this model can be utilized in the initial assessment or case conceptualization phase of treatment when utilizing CBT. The aim of the model is first to make the client aware of how any of the five aspects can impact on each other: 'They all sit inside our environment and interact with it as well. What we feel is closely connected to our thinking, our behaviour, our biology and our environment' (Padesky & Mooney, 1990, p. 13). Secondly, the model aims to help the client understand how changing one of the aspects can impact on the others getting 'better'. In Figure 11.1, this model is illustrated with the hot thought elicited by TJ.

Possible phobia

A phobia is an irrational, intense, persistent fear of certain situations, objects, activities or persons. The main symptom of this disorder is the excessive, unreasonable desire to avoid the feared subject (Bourne, 2005).

During the first session TJ expressed concerns about how she was currently reacting to watching rugby at matches or on television. TJ reported that during a team

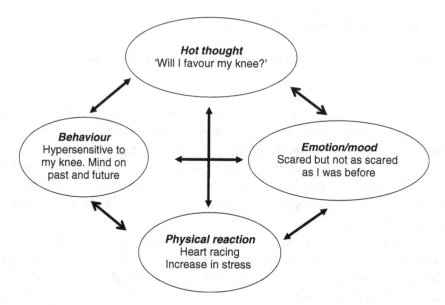

Figure 11.1   Five aspects model (Padesky and Greenberger, 1995) with 'hot thought'.

meeting she had watched a replay of the match in which her injury was replayed in slow motion. She also recounted how she had found herself wincing when she was watching another game on television and reported that this was very unusual for her. This concerned her for a number of reasons. Firstly, if this was her reaction to seeing games on television, she worried about how she would react when she was playing. Of lesser importance to TJ, but still pertinent, was the fact that she utilized observation of rugby on television as a way to study the game, to rehearse and learn situational reactions, and to manage her confidence via modelling and vicarious experience (Bandura, 1997). TJ presented evidence indicating she was currently avoiding rugby. Her avoidance included behaviours such as avoiding watching rugby on television, avoiding attending live rugby matches and avoiding rugby related imagery. Following the initial case conceptualization a case conference was scheduled with myself, the physiotherapist, and the strength and conditioning coach. Based on the initial needs assessment three objectives were set from a psychological perspective:

1. A CBT intervention to be implemented to optimize the rehabilitation process.

2. A systematic desensitization intervention to be established and monitored.

3. The multi-disciplinary team to be communicating and working effectively.

11.3 Interventions and monitoring

Over a three month period TJ had 21 physiotherapy sessions, 20 strength and conditioning sessions and 10 psychology sessions (some of these included joint sessions).

A CBT intervention to be implemented to optimize the rehabilitation process

A CBT intervention was designed to help TJ manage her interpretation of her experience. The underlying premise of CBT is that if you want to change the way you feel, you need to change the way you think (Padesky & Greenberg, 1995). It has been suggested that key parts two effective rehabilitation is exploring the cognitions and perceptions the athlete has of themselves and their injury early in the rehabilitation process (Andersen, 2005) followed by challenging incorrect or naive assumptions about the injury (Moran, 2004).

The aim of the intervention was for TJ to have the ability to recognize and challenge negative thoughts and to generate alternative thoughts. This would help TJ to focus on external factors specific to rehabilitation and her strength and conditioning work, instead of focusing on negative automatic thoughts that may have an internal locus of attention and could also lead to negative changes in mood, behaviour and physical reactions (Ives & Shelley, 2003).

In this CBT intervention the following steps were utilized:

- initial and ongoing case conceptualization – an education process for both the practitioners and the athlete;

- identifying negative automatic thoughts, rating moods and finding supporting evidence;

- identifying alternative thoughts and finding supporting evidence;

- ongoing checking and adjusting for faulty thinking patterns and biased thinking.

To recap, the initial hot thought for TJ was 'will I favour my knee?', and was linked to her mood of feeling scared. The next step was to investigate evidence for and against this thought, and then to generate alternative thoughts. In Figure 11.2 we can see that the alternative thought that TJ decided on in this instance was 'fantastic pins'.

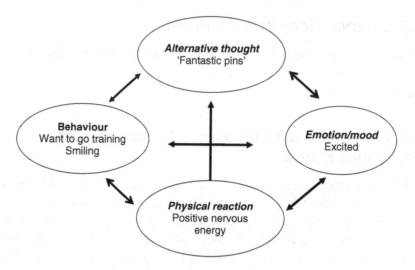

Figure 11.2 Alternative thought.

As a framework for investigating evidence, a thought chart (Padesky & Greenberger, 1995) was utilized. During individual and joint sessions (i.e. with other members of the support team present), a review of automatic thoughts was completed, and in addition, any thoughts that emerged during the session were addressed using the cognitive behavioural therapy skills highlighted in the thought record. An example of this was during one joint session early in the rehabilitation process. Both TJ and the physiotherapist were referring to her right and left knee respectively as good old bad knee and new bad knee. By using socratic questioning and guided discovery (Padesky, 1993) I helped the athlete explore how these thoughts led to unhelpful emotions, behaviours and physical reactions and whether the focus could be on something more productive, namely the thought of 'fantastic pins'.

After each session 'homework' was set in terms of what TJ wanted to achieve by the next session. Care was taken that the goals set were compatible with the physiotherapy and strength and conditioning goals. Goal setting was completed in conjunction with the other practitioners during joint sessions. When the psychology sessions were held independently, permission was sought from TJ to share the goals with the rest of the support team.

A systematic desensitization intervention to be established and monitored

To recap, TJ was concerned that she had winced when watching rugby on television. This had made her wonder how she would react when she saw live rugby, and more importantly, how she would feel about actually playing. A schedule of systematic desensitization (Wolpe, 1958) was developed in conjunction with TJ to enable graded exposure to behaviours she had designated as potentially problematic. This was done in conjunction with the CBT work explained above. Systematic desensitization is a form of behavioural therapy used to effectively overcome phobias, utilizing relaxation skills in conjunction with graded exposure to a hierarchy of fears. The idea is that the individual will be able to progressively cope with increasingly graded fears. Research has shown that a combination of CBT and systematic desensitization can prove effective in treating phobias (Beck, Emery & Greenberg, 1985; Kennerley, 1997). TJ's systematic desensitization schedule is shown in Figure 11.3 alongside the relaxation strategy she chose to utilize.

The multi-disciplinary team to be communicating and working effectively

Multi-disciplinary teams are seen as a productive approach to working in sport science support (Rowell, 1998). It has been suggested that helping to develop constructive perceptions of injury and rehabilitation is best done as part of a team and not in isolation (Moran, 2004). A collaborative approach to management of perceptions in the rehabilitation process involving all the physical and psychological specialists may be the key to effective rehabilitation (Moran, 2004).

One potential hurdle in multi-disciplinary teams is the impact of confidentiality (Collins, Moore, Mitchell & Alpress, 1999). Within a support team of doctors, physiotherapists, strength and conditioning coaches, psychologists and nutritionists there will be different professional codes of conduct. These codes of conduct will have different perspectives on the role of confidentiality. During the first session with TJ, her rights to confidentiality were explained and permission was requested, and granted, for information to be disclosed to designated colleagues in order to provide the most effective service. This was in accordance with the code of conduct of the British Psychological Society which makes provision for the limitations of maintaining confidentiality in multi-disciplinary teams. Under point 1.2 Vb in terms of standards of privacy and confidentiality the code states the following:

1.2 Standard of Privacy and Confidentiality.

(V) Ensure from the first contact that clients are aware of the limitations of maintaining confidentiality, with specific reference to:

(b) the likelihood that consultation with colleagues may occur in order to enhance the effectiveness of service provision;

Throughout the intervention period TJ's permission was requested for attendance at joint sessions and for sharing of specific information.

Systematic desensitization hierarchy

1. Watching videos of Canada/South Africa last matches.

2. Watching television/videos of live matches.

3. Observing a live match.

4. Doing imagery of hitting the machine straight.

5. Hitting the machine straight.

6. Doing imagery of hitting someone holding a bag.

7. Hitting someone holding a bag.

8. Doing imagery of full contact practice.

9. Full contact practice.

Relaxation strategy: the steps

1. Stop the footage.

2. Relax deeply – use the breathing that Raph taught you and practice your pre-match smile.

3. Rewind and start the tape again. This may cause you to tense up somewhat.

4. Begin immediately to regain your relaxed state using the breathing that Raph taught you and practice your pre-match smile.

5. As you do allow your relaxation to deepen.

6. When you can watch the bit of the match which made you wince without any discomfort go on to the next point on your hierarchy.

Figure 11.3 Systematic desensitization hierarchy and relaxation steps.

Other potential stumbling blocks for working in multi-disciplinary support teams in sport were highlighted by Reid, Stewart and Thorne (2004) who applied Aamodt's (1999) structural risk factors for conflict escalation. The team sought to consider these factors and limit their impact on their effectiveness when intervening, as highlighted below:

Competition for resources is present in the field of sports, as future resourcing is often dependent on perceptions of service efficacy by clients or service providers. Competition for resources was not apparent as all practitioners worked within the English Institute of Sport. Therefore there was no direct marketing or bidding for future resources. Nonetheless, this athlete was a significant link to athletes and coaches within the RFUW so future acceptance, integration and potential work was dependent on a professional intervention.

Task interdependence and jurisdictional ambiguity are apparent in the sports world in which numerous practitioners may be providing differing opinions on issues and in which practitioners may be blurring the lines of their professional boundaries (Reid *et al.*, 2004). This was not an issue with this team as there was a culture of openness among the practitioners and a desire to learn about the other disciplines. This learning environment was especially apparent in joint sessions, where there was an eagerness to understand more about each other's disciplines and to help each other develop as practitioners. There was also clarity on professional boundaries due to prior work delivered with the whole EIS London team on boundaries of competence and practice, and role clarity.

Communication barriers may hinder effective sharing of key information and may result in only key personnel being privy to knowledge. In addition, sporting practitioners who are 'experts' in their own field may be overconfident and may struggle to be open to the opinions and perspectives of others. Within the learning environment developed in the EIS the practitioners were able to be open to constructive criticism from each other, therefore communication was not a significant barrier. The three practitioners met informally on a daily basis and would 'check in' with each other when we were due to meet TJ to share any pertinent information relevant to their session. Indeed the positive and constructive feedback within this group of practitioners may have led to the whole being greater than the sum of its parts, a key component of system theory identified by Berg-Cross (2000).

Group homogeneity and group size can also impact on the team effectiveness. Groups of five have been identified as having the ideal mix of commitment and collaboration. Getting the balance between homogeneity and heterogeneity is another key factor. In terms of proximity all three practitioners were based in the same work place five days a week. The size of the team was not the optimum five suggested by Reid *et al.* (2004), but sufficiently large for heterogeneity to be explored.

The ability of the multi-disciplinary team to work effectively can be illustrated by the collective management of perceptions (Moran, 2004) and the effectiveness of the integrated goal setting.

Management of perceptions

The support provided to TJ by the practitioners included the reframing of biased thinking. An example of this was the use of the Quadometer. The Quadometer, an idea generated by the physiotherapist, was a measurement tape hung in the physiotherapy room. The goal was to see how big TJ's quadriceps could become during the rehabilitation period. If TJ expressed biased thinking such as 'will this be 2002 again?', then collectively the practitioners helped the management of this perception, often in joint sessions, by helping her reframe to the thought 'Quadometer'. Another example of this was in the use of imagery during the rehabilitation process. A key image that TJ utilized during her rehabilitation was picturing how her muscles had knitted together in her knee. This type of imagery has been advocated as part of rehabilitation by a number of researchers (Driedeger, Hall & Callow, 2006; Milne, Hall & Forwell, 2004; Sordoni, Hall & Forwell, 2002). The recall of this image, used as another approach to reframing, was reinforced by all three practitioners to help support TJ.

Integrated goal setting

The integrated goal setting enabled the team to be 'singing from the same hymn sheet'. At the first session with me, TJ set goals relating to how she wanted to perform at the World Cup which at the time was 11 months ahead. Her goals were to:

- Retire having been the best second row at the World Cup.

- Help the team to perform to their best by leading by example with regards to intensity and aggression.

To enable effective goal setting across disciplines role clarity and acceptance was paramount (Reid et al., 2004). The decision on the type of goal setting that would be suitable for TJ was based on research (Weinberg, Burton, Yukelson & Weigand, 2000; Gianni, Weinberg & Jackson, 1998) which has suggested the importance of successfully matching goal orientations and goal preferences to maximize the impact on performance. Weinberg et al. (2000) found that Olympic athletes rated overall performance, winning and having fun/enjoyment as the most important goals. Those athletes who were win-orientated felt that competitive outcome goals were more effective for them, whilst those who were more performance-orientated found that focusing on goals that improved their physical and psychological state were more

Manage good execution and performance in the lineout

- Good jump and catch from me.
- Good calls (analyse their pack's strengths, review with our pack night before, calls on wristband).
- Adjustments/clarifications/motivation of pack.

Make tackles

- Positive thoughts.
- 'Safer' positioning – guard and body guard.
- Lower body height.
- 'Smash' – key word.

Effective involvements/few mistakes

- Pick and choose involvements carefully.
- Don't hesitate on decisions (go with instinct).
- 'Go for it' – key words.

Figure 11.4 TJ's goals for trial match.

beneficial. To recap, the goals TJ set herself were predominantly performance-based. These goals were shared amongst the support team and were then integrated into the rehabilitation goals. To integrate the goals it was agreed amongst all three practitioners that for TJ detailed planning and goal setting would be a key factor in her effective rehabilitation. There was recognition amongst the practitioners of the need for precise planning in the strength and conditioning and physiotherapy arena as TJ responded well to planning followed by action.

In addition, as TJ approached key milestones, goals were generated by asking TJ to explore how she was going to judge if she was successful, in order to promote a focus on performance goals. This led to specific, motivational goals which meant TJ had a clear picture of what she wanted to achieve. This goal setting also helped prepare TJ to be in the correct frame of mind for each new milestone. An example of this can be seen in the goals TJ set to judge if she was successful at the trial match, which are illustrated in Figure 11.4.

11.4 Evaluation of intervention

In order to evaluate the effectiveness of the psychological intervention, it is important to evaluate the three objectives set at the start of the intervention.

A CBT intervention to be implemented to optimize the rehabilitation process

To assess whether this objective was achieved, it is worth considering what behaviours and thinking styles TJ would display if the CBT intervention was successful. Adherence to her programme would be one of the behaviours (Ives & Shelley, 2003) and TJ's adherence to her rehabilitation and strength and condition programme was exemplary. She was also highly motivated, another key behaviour highlighted by Ives and Shelley (2003). In terms of her progression, she reached all the targets set by the physiotherapist and strength and conditioning coach and also managed to maintain her fitness scores at official squad testing in spite of her time away from training.

In terms of TJ's ability to reframe her thoughts over the intervention period my observations and record keeping illustrate that she was able to do this initially with prompting and guided discovery, and over time was able to reframe quickly and independently. Reports from the strength and conditioning coach and the physiotherapist and observations by myself were that TJ quickly shifted her focus away from the hot thought of 'will I favour my knee?' to focusing on positive alternative thoughts such as 'fantastic pins', 'the Quadometer' and 'the muscles knitted together'. Over time she also enhanced her ability to reframe and was thus able to identify and alter thinking biases which naturally occurred during her rehabilitation.

A systematic desensitization intervention to be established and monitored

TJ informally reported that, utilizing her previously learnt relaxation skills, she was able to move up the hierarchy and was soon talking openly about matches she had watched on television and progressed on to observing live matches and then taking part when appropriate. TJ was therefore able to utilize one of her key preparation skills before matches, namely watching video and television footage of rugby matches, imagining herself in similar situations and using vicarious experience as a mechanism to enhance her self confidence (Bandura, 1997).

The multi-disciplinary team to be communicating and working effectively

This objective was successfully achieved. The multi-disciplinary team worked very effectively together, as is illustrated in the quote below from the strength and conditioning coach.

> When high quality practitioners working regularly with each other – so communication is easy – working with a highly motivated, naturally positive – optimistic athlete – who drives things forward and does everything you tell them to the letter – things tend to go quite well!!

Further evaluations from the strength and conditioning coach illustrate how the goals set by the strength and conditioning coach integrated with the physiotherapy guidelines and were based around the cognitive state that TJ and the team had assessed as optimal for her in terms of managing her perceptions:

> Even when TJ was in early rehabilitation she and I set positive goals to achieve in training that wouldn't affect her knee, e.g. upper body strength/mass, rear shoulder muscle balance. Also goals/measures for rehab process were clear and well planned. In addition each mini benchmark was positively reinforced. Each mini setback was dealt with rationally.

It would seem that the multi-disciplinary intervention, in conjunction with the quality of the sport medicine in terms of the surgery and guidance from the doctors, played a significant role in TJ's rehabilitation.

To examine the full impact it is worth reflecting on what TJ did next. TJ, at the end of her rehabilitation, was a fully fit and functioning world class athlete as measured by fitness testing, performances at Six Nations trials and selection into that squad. She was then a key player in a successful Six Nations campaign in early 2006. She arrived at the World Cup in September 2006 fully fit. She was in the starting 15 and played the whole match in the 2006 World Cup Final. Her performances at the World Cup were honoured by her being voted into IRB Women's World XV team thereby fulfilling one of TJ's key goals set at the outset of the rehabilitation, to be rated the best second row at the World Cup.

11.5 Evaluation of consultant effectiveness/reflective practice

To evaluate and reflect on the consultant effectiveness and, in this case, the effectiveness of the support team, a reflective practice model was utilized as a framework (see Gibbs, 1988). The reflective practice cycle begins with 'What Happened During The Experience?' and relates to factual occurrences that have been covered previously in this chapter.

What were you thinking and feeling during this experience?

My reflections on this experience is of a high performing team working together towards a clear goal with a highly motivated and professional athlete. I remember thinking that I needed to be 'on top of my game' during both individual and joint sessions as I was aware of the high calibre of people I was working with. I felt highly motivated being part of this team and knew that I had a role to play in a number of areas. Initially, it was to help TJ deal with a possible phobia she had about rugby and returning to rugby. I also felt I needed to highlight the importance of appropriate goal setting to the other practitioners. This was not in terms of educating them as they were highly skilled practitioners, but instead to give them a psychological perspective on this area. It was also key that we helped TJ set appropriate measures of success at different times during her rehabilitation. For example, it was important that she set the right types of goals for her first return to squad training, for the Six Nations trial and for her first match back. These were important transitional periods and I think we helped give her clear direction about how to judge if she was successful.

What was good and not so good?

I felt that we managed the area of confidentiality in accordance with our codes of conduct. I also believe that the interaction between the support team was excellent. The team seemed able to motivate, challenge and support a highly motivated professional athlete. There was also an excellent relationship with the RFUW head coach and doctor.

The interventions were effective tools in altering the thinking patterns of the athlete and helping her deal with anxieties about returning to rugby. The goal setting intervention was effective and this may have been in part due to the successful matching of goal preferences which has been identified as having an impact on performance and effort (Weinberg *et al.*, 2000; Gianni *et al.*, 1998). Overall the work of the multi-disciplinary team addressed effective rapport building, managing perceptions of the situation and maintaining motivation through effective goal setting, factors which Petitpas (2002) has identified as critical in the rehabilitation of injury.

In terms of what was not so good, it would have been beneficial *at the time* to do more formal reflections on the *process* of support. This could have involved formalized feedback between practitioners and filming of individual and joint sessions to help uncover processes that may have been forgotten or overlooked. TJ had been affected by watching the footage of her injury in slow motion after the incident. With hindsight, it may have been beneficial to manage her exposure to such footage at that time.

What sense can you make of this?

In reflecting on what sense I could make of this highly successful multi-disciplinary intervention, I looked to various research to understand why the group had worked so effectively. There is considerable research evidence supporting the impact of CBT, goal setting, motivational strategies and systematic desensitization. What I felt I needed to make more sense of was how the support team worked effectively.

One key reflection was how the strength and conditioning coach became the key stakeholder in the functional effectiveness of the group. The literature on transformational and transactional leadership (Podsakoff, Mackenzie, Moorman & Fetter, 1990) provided me with a clear picture of how he played this informal leadership role so effectively.

Transactional leadership is viewed as a traditional way to lead in which good performances are rewarded and current work processes are maintained as long as performance targets are being met (Arnold, Coopers & Robinson, 1998). This type of leadership is viewed as suitable for stable work environments in which bureaucratic processes prove effective (Bass, 1990).

Transformational leadership aims to develop, inspire and challenge employees to enable them to stretch beyond their own self-interest and buy into a higher collective vision (Arnold et al., 1998). This type of leadership is viewed as ideal for changing environments (McKenna, 1994; Tichy & Devanna, 1986), and the focus is on leaders with vision, innovation and creativity who are able to galvanize others to follow their dreams.

This case study occurred in a changing environment, so having a transformational leader was ideal. Two of the practitioners involved had only been in post for two months with the English Institute of Sport. The athlete was also in an injured state, this itself creating a changing environment. Lastly, the three practitioners were in a novel environment of all three having to interact and work with each other in a full-time capacity. The environment itself had changed for the 'leader' (the strength and conditioning coach), for up to this point he had not worked in this specific rugby injury environment with a psychologist and a physiotherapist. In hindsight there was a lot of potential for 'stepping on toes' (competition for resources), which has been highlighted as a risk factor in effective multi-disciplinary sports teams (Reid et al., 2004).

Podsakoff et al. (1990) have shown that six transformational leadership behaviours (identifying and articulating a vision, providing an appropriate model, fostering the acceptance of group goals, high performance expectations, individualized support and intellectual stimulation) and one transactional leadership behaviour (contingent reward behaviour) were able to predict employees' attitudes to their work, as demonstrated by job satisfaction, organizational commitment and trust in, and loyalty to, the leaders. This ability of transformational leaders to have an impact on people's attitude has been reported elsewhere in the literature on effective teams (Russell, 2001).

Within the multi-disciplinary team that supported TJ there was no employer or employees. Nonetheless there was a clear leader in the team, the strength and conditioning coach. It is worth at this junction exploring how he developed into this role before returning to see how his transformational leadership impacted on his colleagues and the athlete.

Reviewing the history of the three practitioners' involvement with England Women's Rugby, there are striking differences in terms of length and immersion in the sport. The strength and conditioning coach had been the lead strength and conditioning coach for the RFUW since 2003, had travelled with the team and was part of the World

Cup squad. The physiotherapist and I had been working with RFUW since 2005 and did not travel with the squad. The strength and conditioning coach had an excellent and long-established working relationship with the head coach and the support staff. Therefore he had, because of this experience, a large amount of specific contextual intelligence (Koskie & Freeze, 2000). This resulted in him being able to act as a powerful driver of interventions and change.

The first transformational behaviour to examine is *identifying and articulating a vision*. The vision was painted in conjunction with TJ and revolved around creating a highly motivating, stimulating and challenging environment around a highly professional athlete. In terms of *providing an appropriate model*, this was done using the framework proposed by Ives and Shelley (2003) for effective functional strength and power training based on psychophysics. This enabled the athlete to have the appropriate cognitive state to positively influence her physiological adaptations to the strength and conditioning training programme. The third transformational behaviour of *fostering the acceptance of group goals* involved high energy and vision from the strength and conditioning coach and the highly motivated professional attitudes of the other practitioners, which led to a high level of 'buy in'. In terms of *high performance expectations*, there was an onus placed within the team on 'delivering'. In addition there was a drive from within the EIS to have a measurable impact on performance. This can be seen in the quote from the strength and conditioning coach below:

> The confidence in the other practitioners' ability to deliver was vital as the athlete was driving a tight timetable. This confidence stemmed from the positive operation outcomes, the knowledge of a top knee surgeon and the trust and confidence the team had in each other to deliver.

Individualized support was demonstrated by the strength and conditioning coach through his willingness to understand and implement ideas from different specialists. This linked into the final transformation behaviour of *intellectual stimulation* with the strength and conditioning coach actively soliciting ideas and input from both the physiotherapist and the psychologist to enhance his own work.

The last behaviour that Podsakoff *et al.* (1990) highlight as having an impact on employees' attitudes, and thus their subsequent behaviour, is the transactional leadership behaviour of *contingent reward*. Contingent reward denotes the establishment of appropriate rewards contingent on the reaching of clarified expectations. In this case study no formal expectations were set by the strength and conditioning coach, but informal expectations were placed on the practitioners by themselves and there was a high degree of informal positive feedback on the quality of the work done by the practitioners. This happened across the practitioners but, on reflection, the impact of the feedback from the strength and conditioning coach due to his key stakeholder role may have had added value. Personally, in this type of environment and with a motivational informal leader I was able to work to a world class standard in a challenging, educating and enjoyable environment.

What else could you have done in the situation?

I feel we could have had formal reflective practice of our work. We could also have asked TJ for formative feedback during her rehabilitation and feedback at the end of the intervention to impact on current and future work. This type of reviewing may have highlighted key aspects in the communication style of the practitioners and identified elements of consultant effectiveness that are increasingly a focus in applied sport psychology research (Anderson, Miles, Robinson & Mahoney, 2004; Tod & Andersen, 2005).

If a similar experience arose again, what would you do?

At the start of the process I would ascertain the specific goals of the intervention, clarify the informal and formal roles of the multi-disciplinary team, discuss the type of leadership required and reinforce the importance of effective communication. I may utilize psychometric tools to develop a richer understanding of the athlete, and that may also enhance practitioner communication. Finally, I would utilize a more structured form of reflective practice during the process.

11.6 Summary

This chapter has focused on the use of a cognitive–behavioural therapy intervention for injury rehabilitation in an international rugby union player. The key processes necessary for effective multi-disciplinary work between psychologist, physiotherapist and strength and conditioning coach have been highlighted. The chapter shows the role that a sport psychologist can play in a multi-disciplinary team and the structures and behaviours that need to be in place for such teams to flourish. The cognitive–behavioural therapy intervention involved the identification, challenging and altering of thinking patterns. By the end of the intervention the athlete was autonomous in her ability to challenge and change her thoughts. A systematic desensitization intervention also resulted in the athlete being able to address and manage fears about watching and playing rugby after injury. Alongside these specific interventions, the importance of processes relating to transformational and transactional leadership was identified, and risk factors for effective functioning of multi-disciplinary support teams were actively managed.

Questions for students

1 Why is a cognitive–behavioural therapy approach suitable for working with an injured athlete?

2 What types of goals is it important to set during rehabilitation from injury?

3 What are the potential areas of conflict within a multi-disciplinary team?

4 What elements of applied practice made this multi-disciplinary team effective?

5 What is the importance of confidentiality in the practice of sport psychology? How can this be managed in a multi-disciplinary support team?

References

Aamodt, M. (1999) *Applied Industrial/Organizational Psychology*. Wadsworth, Belmont, CA.

Andersen, M.B. (2005) *Sport Psychology in Practice*. Human Kinetics, Champaign, IL.

Anderson, A., Miles, A., Robinson, P. *et al.* (2004) Evaluating the athlete's perception of the sport psychologist's effectiveness: what should we be assessing? *Psychology of Sport and Exercise* 5, 255–277.

Arnold, J., Cooper, C. L. and Robinson, I. T. (1998) *Work Psychology: Understanding Behaviour in the Workplace*. Prentice-Hall, London.

Bass, B. M. (1990) From transactional to transformational leadership. *Organizational Dynamics* 18, 19–31.

Bandura, A. (1997) *Self-Efficacy: the Exercise of Control*. W.H. Freeman, New York.

Beck, A.T., Emery, G. and Greenberg, R.L. (1985) *Anxiety Disorders and Phobias: A Cognitive Perspective*. Basic Books, New York.

Berg-Cross, L. (2000) *Basic Concepts in Family Therapy: An Introductory Text*. Haworth Press, New York.

Bourne, E.J. (2005) *The Anxiety and Phobia Workbook*. New Harbinger Publications.

Brewer, B.W. (1994) Review and critique of models of psychological adjustment to athletic injury. *Journal of Applied Sport Psychology* 10, 1–4.

Collins, D., Moore, P., Mitchell, D. *et al.* (1999) Role conflict and confidentiality in 16 multidisciplinary athlete support programmes. *British Journal of Sports Medicine* 33, 208–211.

Driedeger, M., Hall, C. and Callow, N. (2006) Imagery use by injured athletes: a qualitative analysis. *Journal of Sports Sciences* 24, 261–271.

Giannini, J., Weinberg, R.S. and Jackson, A. (1998) Effects of mastery, competitive and cooperative goals in the performance of complex and simple basketball skills. *Journal of Sport Psychology* **10**, 408–417.

Gibbs, G. (1988) *Learning by Doing: A Guide to Teaching and Learning Methods*. Oxford Brookes University, Further Education Unit, Oxford.

Ives, J.C. and Shelley, G.A. (2003) Psychophysics in functional strength and power training: review and implementation framework. *Journal of Strength and Conditioning Research* **17**, 177–186.

Kennerley, H. (1997) *Managing Anxiety: A Training Manual*. Oxford University Press, New York.

Koskie, J. and Freeze, R. (2000). A critique of multidisciplinary teaming: problems and possibilities. *Developmental Disabilities Bulletin* **28**, 1–15.

McKenna, E. (1994) *Business Psychology and Organizational Behaviour*. Erlbaum, Hove.

Milne, M., Hall, C. and Forwell, L. (2004) Self-efficacy, imagery use, and adherence to rehabilitation by injured athletes. *Journal of Sport Rehabilitation* **14**, 150–167.

Moran, A.P. (2004) *Sport and Exercise Psychology: A Critical Introduction*. Routledge, Hove.

Padesky, C.A. (1993). Socratic questioning: changing minds or guiding discoveries? Keynote address presented at the meeting of the European Congress of Behaviorual and Cognitive Therapies, London.

Padesky, C.A. and Geenberger, D. (1995) *Clinician's Guide to Mind Over Mood*. The Guildford Press, New York.

Padesky, C.A. and Mooney, K.A. (1990) Clinical tip presenting the cognitive model to clients. *International Cognitive Therapy Newsletter* **6**, 13–14.

Persons, J. (1989) *Cognitive Therapy in Practice: A Case Formulation Approach*. W.W. Norton, New York.

Petitpas, A.J. (2002) Counseling interventions in applied sport psychology. In: Van Raalte, J.L. and Brewer, B.W. (eds), *Exploring Sport and Exercise Psychology*, pp. 253–268. American Psychological Association, Washington, DC.

Podsakoff, P. M., Mackenzie, S. B., Moorman, R. H *et al.* (1990). Transformational leader behaviours and their effects on followers' trust in leader, satisfaction, and organizational citizenship behaviors. *Leadership Quarterly* **1**, 107–142.

Reid, C., Stewart, E. and Thorne, G. (2004) Multidisciplinary sport science teams in elite sport: comprehensive servicing or conflict and confusion? *The Sport Psychologist* **18**, 204–217.

Rowell, S. (1998) *Sport Science Support Programme Review*. National Coaching Foundation, Leeds.

Russell, C. J. (2001) A longitudinal study of top-level executive performance. *Journal of Applied Psychology* **86**, 560–573.

Sordoni, C., Hall, C. and Forwell, L. (2002) The use of imagery in athletic injury rehabilitation and its relationship to self-efficacy. *Physiotherapy Canada* Summer, 177–185.

Tichy, N. M. and Devanna, M. A. (1986) *The Transformational Leader*. John Wiley & Sons, New York.

Tod, D. and Andersen, M. (2005). Success in sport psych: effective sport psychologists. In: Murphy, S. (ed.), *The Sport Psych Handbook*, pp. 305–314. Human Kinetics, Champaign, IL.

Udry, E. and Andersen, M.B. (2002). Athletic injury and sport behaviour. In: Horn, T.S. (ed.), *Advances in Sport Psychology*, pp. 529–553. Human Kinetics, Champaign, IL.

Weinberg, R., Burton, D., Yukelson, D. *et al.* (2000) Perceived goal setting practices of Olympic athletes: an exploratory investigation. *The Sport Psychologist* **14**, 279–295.

Wiese-Bjornstal, D.M., Smith, A.M., Shaffer, S.M. *et al.* (1998) An integrated model of response to sport injury: psychological and sociological dynamics. *Journal of Applied Sport Psychology* **10**, 46–69.

Wolpe, J. (1958) *Psychotherapy by Reciprocal Inhibition*. Stanford University Press, Stanford, CA.

12

Developing Coach Education Materials in Table Tennis – Applying a Cyclical Model of Performance

Tim Holder

St Mary's University College, London, UK

12.1 Introduction/background information

In 1999 a meeting was initiated by the then Director of Coaching and Learning with the author to discuss a different approach to aspects of the coach development programme within the sport of table tennis. The fundamental basis of this discussion was to establish a framework within which the development of coaches' understanding of motor learning and psychological skills training could be organized. At the time of the meeting, I had 10 years' experience within the sport, working as an applied sport psychologist in the delivery of both performance enhancement strategies with elite performers at World and Olympic events and working within the coach education scheme delivering sport psychology materials. This delivery was enhanced through utilizing experiences within the sport to make the materials maximally relevant to the specific coaching audience. In addition to this, the area of my doctoral research had been in anticipation and skill acquisition within table tennis, much of which was of

Applied Sport Psychology Edited by Brian Hemmings and Tim Holder
© 2009 John Wiley & Sons, Ltd

direct relevance to coaches and performers. The sport had a history of being open to new ideas and approaches and had been involved in the funded application of sport science since its inception in the late 1980s (originally through the UK Sport Science Support Programme).

The nature of the discussion between myself and the Director of Coaching and Learning reflected approaches that were different in theoretical emphasis but complementary in their applied intent (i.e. focused on identifying ways of helping coaches and performers to be more effective). The Director of Coaching and Learning clearly identified the sport-specific demands and ensuing challenges for coaches in helping performers, and possessed considerable experiential expertise. My perspective brought a more scientific and applied research-based input to the discussion. This was based on available evidence, and a rapidly developing understanding of how sport skills are learnt and retained, and the impact of psychological skills on performance (e.g. Williams, 2006).

Because of this mix of expertise, and the mutual respect and trust that had been established over a number of years of contact, the process of discussion enabled the formation of an underpinning framework in the form of the cyclical model presented later in this chapter.

The role of the sport psychologist

It is clear that the majority of literature within the area of applied sport psychology alludes to the key role of practitioners in developing mental skills within performers for the purpose of performance enhancement (e.g. Murphy, 2005). This emphasis is appropriately reflected within this book. What is rarely identified is the role of the applied sport psychologist in working with a skill acquisition and coach education emphasis. Additionally, the incorporation of psychological skills into the *training* context has been regarded as an opportunity to practise and develop the skills to enhance transfer into the competitive environment. Alternatively, incorporating such psychological skills can be specifically used to enhance skill development (see Sinclair & Sinclair, 1995). The emphasis towards competition preparation downplays the role which psychological skills can play in interacting with the challenges of a *training* environment. For example, the use of imagery as a technique to enhance confidence (Callow & Hardy, 2001) can be used very effectively within a training context to aid the performer in maximizing the impact of training time and help maintain motivation in the face of failure. The integration of psychological skills into the training context is familiar to coaches and performers, but is rarely (in my experience) used in a systematic manner. Therefore the role of the applied sport psychologist in this context extended beyond the more familiar performance enhancement/competition preparation focus into the realms of integration and utilization of psychological skills in a skill-learning context. Furthermore, the development of coach education materials, and the design of delivery approaches for this information, became a significant challenge to maximize use of the limited time available. Within the coach education and training scheme, 6 h were allocated to developing an understanding of the practical coaching implications of the information in skill acquisition and psychological skills training.

Motor learning approaches

The working practices of applied sport psychologists can vary significantly based upon their individual training, subsequent expertise and resulting confidence in working in particular contexts. As a result of this, many practitioners consider themselves to be more highly skilled in the mental training and social psychological aspects of the field. In contrast to this, my own educational training and research are diverse and include social psychology and applied sport psychology as well as skill acquisition frameworks. The juncture at which such training becomes optimally beneficial when working as a sport psychologist is clearly when working within the realms of coach education and coach development. The primary focus of this work tends to be on skill acquisition and training.

The main motor learning frameworks used within sport are the information-processing approach (Schmidt & Wrisberg, 2000) and the ecological framework (Davids, Handford & Williams, 1994). The ecological perspective has more recently been developed into a 'constraints-led' approach that has gained significant momentum in terms of both research evidence and materials aimed at converting such research into effective coaching strategies (e.g. Davids, Button & Bennett, 2008). The more traditional information-processing framework emanates from a cognitive perspective whilst the ecological and constraints-led approaches can be aligned more closely to a behaviourist framework. It is clear that both these approaches have something to offer an applied sport psychology practitioner whose framework is a cognitive–behavioural one. Thorough examination of these perspectives is beyond the scope of this chapter. However, a summary of key theoretical characteristics is necessary to indicate which factors influenced the work completed.

The information-processing framework depends upon a memorial representation as a central conceptual starting point for how performers can improve their skills and, ultimately, performance. The development and elaboration of schema (Schmidt, 1975), or other alternative formulations of generalized motor programmes, forms the platform from which coaching practice can be enhanced. The development of effective, memory-stored, movement commands is purported to be influenced by a range of factors such as practice structure, feedback and attentional control (Vealey, 2005).

The ecological framework provides a shift away from internally driven, cognitively based assumptions that lead the information-processing approaches. In contrast, the ecological approach focuses on the utilization of information available in a sporting context. This information can be incorporated into action 'on line' without the requirement for comparison to previous experiences (e.g. Bootsma & van Wierengen, 1990). The use of the information is thought to occur through a close coupling, developed through experience, between what is perceived and the action itself. This is a more dynamic conceptualization of movement control and skill development that is at odds with the computer metaphor explanations of the information-processing framework. However, what this approach offers, which is of central importance in most interactive sports, is an emphasis on picking up essential information sources in order to cope with the demands of the situation.

A third perspective, the constraints-led approach, draws significantly from more ecological explanations of movement control. The key premise, from the original work of Newell (1985), is that co-ordinated and controlled movement is generated in a dynamic and context-specific manner by the constraints within that situation. Three broad categories of constraints are identified: organismic, environmental and task. Organismic constraints are those related directly to the characteristics of the person performing the movement. For example, the height of the performer will significantly impact on how they co-ordinate their movements, as will their limb length. Environmental constraints could relate to weather conditions or, more likely in a racket sport, lighting and floor conditions. How well the performance context is illuminated and the grip characteristics of the flooring over which the performer must move with speed and precision clearly influence the movements produced (e.g. tennis players often slide into their shots when playing on clay, an option not possible on other surfaces). Lastly, task constraints relate to elements of performance such as the rules of the game and equipment. For example, in table tennis, the characteristics of the bat alter (or constrain) the way in which the movement needs to be co-ordinated to propel the ball over the net and onto the other side of the table. The constraints-led approach sees the interaction of these three categories of constraint as having a fundamental influence on skill development and movement co-ordination. This perspective provides clear recommendations about how manipulation of constraints may be able to enhance the adaptability of movement solutions that may be particularly useful in high-speed, interactive sports such as table tennis.

Discussions, over time, with coaches have demonstrated that they do not profess allegiance to any of the above positions on how skills are learnt and developed. They are more concerned with what they have found to be practically effective. The reflective coach develops, through experience, a range of strategies to influence skill acquisition that they are committed to and are confident of delivering. Coach education opportunities help coaches put their current practice within a theoretical framework that aids understanding of why the methods they employ might be effective. However, a more important potential impact of coach education opportunities is to provide ideas and evidence that challenge coaches to consider whether they are *optimally* effective. My experience, when speaking with many coaches is that, whilst they are open minded to new (or different) approaches, they are confident that the strategies they currently put in place with performers are already effective. They can be concerned over changing anything and often adopt an approach that represents an attitude of 'if it isn't broken don't fix it'. This potential attitudinal barrier from coaches had a significant impact upon the mode of delivery of materials and is explained later.

The development of motor control and learning theory and research, leading to applied understanding and recommendations for coaches, is still coming to fruition in formats available to coaches and performers (e.g. Davids *et al.* 2008). However, this literature is still targeted primarily at undergraduate and postgraduate students of sport science, physical education and other programmes of study, rather than explicitly at coaches. Therefore the sport psychologist's role in this case is to identify, and collaborate in developing, relevant and evidence-based content directed at coaches, with the intent that they will be able to develop knowledge and apply it with performers.

The following chapter traces the ongoing work with a national governing body of sport in providing evidence-based, sport-specific, coach education workshops within the content of coaching qualification courses. Therefore this chapter differs from many presented within this book in two key ways. Firstly, it focuses on work within a national governing body coach education and coach development programme. Secondly, in terms of scientific focus and underpinning theoretical rationale, this case study emphasizes skill acquisition as its primary focus.

12.2 Initial needs assessment

The initial impetus to change the coach education materials had been created from the approach by the then Director of Coaching and Learning. In addition to this, the features of the sport under consideration needed to be accounted for to drive the emphasis of the materials and to create a framework that reflects most appropriately the needs of the sport itself. It is only when the demands of the sport are understood in detail that a clear framework to address the key issues can be effectively explored. This would be significant when working within any sport – if the demands cannot be clearly articulated it is unlikely that the framework adopted, and resulting materials for coach education in skill learning, are going to be effective.

The demands of table tennis

The sport of table tennis has a number of key components that make it a very challenging and exciting sport. Firstly, table tennis is primarily an open sport incorporating a substantial amount of uncertainty and unpredictability. This unpredictability is primarily derived from the interactive nature of the sport. The fact that players have an opponent wishing to do something to them at the same time as they are trying to do something to their opponent, is an essential characteristic (as with all racket sports and most interactive sports). Add to this the speed of the game, with ball velocity in excess of 60 miles per hour, spin characteristics over 150 revolutions per second (Qun, Zhifeng, Shaofa & Enting, 1992) and a relatively small target area to aim at (table dimensions 9 feet by 5 feet), and the magnitude of the demands on performers can be clearly identified. When considered together, this open, unpredictable, interactive, fast ball sport, requiring high levels of precision under excessive time pressure, is a substantial challenge to both players and coaches.

Perceived weaknesses of current coach education and development

In discussion with the Director of Coaching and Learning I was invited to play a part in developing the current framework of coach education. The aim was to develop an alternative emphasis to the materials that addressed more closely the demands of the sport for both players and coaches. In the view of the Director of Coaching and Learning the system in place tended to produce coaches who focused on *technical* development at the expense of *skilful* performers. The evidence to support this view had emerged from observations of coaches, and identification of the strengths and weaknesses of players evolving from the English Table Tennis Association coaching system over a number of years. In an open sport such as table tennis, an inflated emphasis on technical development seemed to ultimately be an unhelpful approach that emphasized style of play over effectiveness. For example, performers may well possess precisely co-ordinated movement patterns and look to have great potential to excel in the sport. However, when under competitive conditions, the same performers may not be as skilled at making accurate, on-table decisions and effectively carrying out tactical plans. As a result, technical precision is unable to reap the expected rewards as the players are rarely able to exert enough control over rallies to enable the technique to be used to their advantage. Skilful performance is required to a greater degree than technical excellence alone.

Therefore in response to this needs analysis, a new focus and guiding model of performance was developed and integrated into the coach education curriculum. The needs analysis therefore established two clear areas for development. Firstly, there was a need for a shift in emphasis towards helping coaches more adequately understand the need to develop skilful performers. Secondly, there was a clear need for the coach education programme to address, with greater focus, the genuine demands of the competitive game of table tennis.

12.3 Intervention and monitoring

As a result of the needs analysis the cyclical model was developed and introduced into the coach education programme in the way described in detail below. The mode of delivery was through 3 hour interactive workshops. These workshops had a focus on firstly developing skilful performers and secondly developing mental toughness. This chapter is focused exclusively on the workshop aimed at developing skilful performers.

Cyclical model of performance

The cyclical model described below (see Figure 12.1) has been based on current understanding of coaching effectiveness (Dorfman, 2003; Vealey, 2005) and models of skill acquisition. The model describes the importance of three different aspects of human functioning that accumulate to enable effective sporting performance. Firstly, the importance of reading the actions of the opponent is identified. Within a fast-ball sport such as table tennis the time demands require performers to be able to anticipate what an opponent is going to do rather than wait to be sure of what is going to happen. Should they wait until bat–ball contact of their opponent, it is unlikely that they will be able to respond with enough speed to counteract the opponent's shot (unless it is played at a very slow pace). The model therefore establishes a fundamental role of perceptual information (reading the opponent). Secondly, the model indicates the subsequent step of deciding on what to do once reading of the opponent has occurred. This step in the cycle will draw upon past experiences of similar situations to help make an appropriate decision in the current circumstances. Lastly, the model identifies the important culmination of the cycle in the production of the physical movement to enable the performer to 'do it' by creating a technical stroke. Should the performer read the opponent effectively, and have accurately made an appropriate decision, then

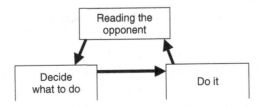

Figure 12.1 A cyclical model of table tennis performance.

the performer has a clear opportunity to make an effective shot in response to the situation. The ability to 'do it' effectively will be limited by the available movement production options developed within the performer through practice.

Therefore the cyclical model clearly describes the need for coaches to develop performers who can not only be technically proficient in their movements (do it) but critically also be able to read what the opponent is about to do and, as a result, make rapid and accurate decisions of what to do (decide what to do). The cyclical nature of the model identifies the importance of *each* of the three stages and how they are mutually constraining. The enhanced emphasis on anticipation (or reading the opponent) borrows much from the ecological approach as well as regarding the development of memorial storage through experience to be key from an information processing perspective.

The emphasis of the model on reading the game takes into account over two decades of research activity within sport science that has established a key element of expertise to be the development of effective anticipation and decision-making (e.g. Abernethy, Wood & Parks, 1999, Baker, Cote & Abernethy, 2003). Key findings of relevance to table tennis have been gleaned from other racket sports and interactive activities (e.g. Shim & Miller, 2003) and incorporate occlusion studies (e.g. Abernethy & Russell, 1987) and visual search data (e.g. Kato & Fukuda, 2002, Ripoll, Kerlirzin, Stien & Reine, 1995) as well as the acknowledgement of research on implicit learning (e.g. Farrow, Chivers, Hardingham & Sachse, 1998) and attentional strategies for learning and performance (e.g. Wulf, McNevin, Fuchs, Ritter & Toole, 2000).

This combination of scientific sources significantly influenced the focus of the content of coach education materials. The content has moved from a central emphasis on technical development to greater consideration of the decision-making and perception elements of table tennis performance.

Content of workshop materials – developing skilful performers

This 3 hour workshop contained the following key content that exemplifies the sport-specific application of the cyclical model:

- A thorough revision of the cyclical model, which the coaches may have been aware of at other stages of coach development qualifications in an introductory manner.

- Discussion of the demands of table tennis. Key demands were highlighted through discussion and linked back to the model. This enabled an exploration of the relative importance of the three aspects of the model – reading the opponent, deciding what to do and doing it (action/technique). The need to anticipate what the opponent is going to do was explored and related to concepts of reaction time (e.g. Welford,

1973), Hick's law (Hick, 1952) and the probabilities of certain shots being played by the opponent (Dillon, Crassini & Abernethy, 1989). The importance of choice reaction time and Hick's law to table tennis is clear. For example, offering opponents lots of potential shot options through playing a weak shot makes it very difficult to respond effectively under the high time constraints of the sport. The probability aspect relates to the cyclical model (Figure 12.1), in that part of anticipating what an opponent is going to do next is inextricably linked to what the performer has just been able to do to them. For example, playing a shot with high speed and spin to one side of the table increases the likelihood of the opponent playing a shot across court (a safer option for the opponent to a high-difficulty shot).

- Implications from the model for practice structure and organization were explored in detail. The level of theoretical background information is limited to those aspects that directly enhance the development of understanding of key features of the model. In light of this, no direct reference to constraints theory (e.g. Davids *et al.* 2008) was included, but implications for practice are clearly driven by both constraints (Newell, 1985) and schema (Schmidt, 1976) approaches. The notion of movement control through programmed actions based in memory and developed to focus on flexibility and adaptability is included. This aspect highlights some key coaching implications. Within table tennis two discrete approaches to practice drills can be identified. They are labelled *regular* and *irregular* practices.

Regular practices are where the players are aware of where and when the ball is going to land in certain parts of the table. The drill will be set up so that one player acts as a feeder and the other player acts as the person upon whom the drill is focused. These regular practices have an enhanced level of predictability inherent to their design and are often used for 'grooving' stroke patterns (technique) and focus more on the action component of the model. The predictable nature of the drill in fact attempts to limit, or even negate, the need for skilful perception of the opponent or the need for 'real world' decision making – with a perfect feeder very low perceptual and decision-making demands are placed on the performer.

Irregular practice is where there is an element of uncertainty within the drill over which the feeder has control. For example the feeder may play the ball to the opponent's forehand side in a regular sequence and have the choice of when they decide to switch the path of the ball to play it to the backhand side of the opponent. Such a drill shifts the demands from the emphasis of regular drills on action to a significant input from reading the opponent and making accurate and speedy decisions. In this form of practice drill reading *each and every shot* that is played by the feeder is essential for optimizing performance on the drill. The necessity to read the opponent makes these drills more like the genuine demands of table tennis.

- The nature of the unpredictable feature used to make the drill more irregular, is an example of where task constraints can be manipulated. For example, elements such as placement of the ball, number of possible alternative shots and the type of shot to be played (e.g. backhand or forehand) are all features of the drill that can be manipulated as part of the task.

- In addition to the notion of regular and irregular practice, the coaches were introduced to the benefits of blocked and random practice schedules for learning skills and for the development of effective, flexible movements (e.g. Goode and Magill, 1986).

- It is clear that some of the coaches were very familiar with performers being successful on regular, blocked practice schedules. They were less familiar with the potential benefits of irregular, random styles of practice. These types of schedules tend to lead to less short-term success but more longer-term effectiveness. The potential dangers of such practice scheduling for players' confidence and motivation was addressed in a further workshop on mental toughness.

- The coaches were also asked to consider the effectiveness of their use of feedback in a training context. The features of feedback discussed were frequency, timing, level of detail and style of delivery to performers (see Wulf & Shea, 2004). It is true that coaches often report using feedback in a variety of ways to aid performers, and the use of available scientific evidence can help them to understand how they may be able to adapt their current practice to enhance its effectiveness.

Style of delivery

Before delivering any of the materials, an important introductory element was presented in order to establish expectations for the coaches of what the psychology-related workshops were aiming to achieve. In this introductory element it was made clear to the coaches that the information provided, and ensuing discussions, would most likely result in one of three potential outcomes:

1. Enable them to 'pat themselves on the back' for how what they do fits into the background theory. This almost always applies to some aspects of the coaches' delivery.

2. Provide explanatory background information that helps them to understand more fully how they are currently effective in their coaching practice.

3. Challenge their current practice, and identify which elements would benefit from adaptation, based on new information that may contradict their current practice. They can begin to establish new ways of dealing with the same situation.

Opening up such possibilities at the start of the workshops provided an expectation that they would learn from each other during the session and resulted in more open debate and peer learning opportunities within the workshops. It also prepared coaches for the possibility that they could be more effective in their coaching without them feeling threatened or that everything they did was in some way 'wrong'.

The style of delivery was primarily through discussion and application to coaching practice. Because of the often diverse groups of coaches encountered on coach education programmes, a significant factor in the successful delivery is to help the coaches develop an understanding of the cyclical model by relating it to their current coaching challenges. Therefore a facilitator role was adopted for these sessions and coach input drove the focus of application. An example of the facilitator approach was an element where coaches were asked to explore the impact of speed and spin for coaching table tennis. The responses from coaches were collated on a flipchart and used as a method for developing awareness and understanding of the cyclical model. Specifically, the demands on the performer were highlighted and discussion of how effectively reading the opponent could help performers to cope with these demands took place.

At the end of the workshop the coaches were asked to identify up to three practical considerations from the sessions that they aimed to incorporate into their future coaching practice.

12.4 Evaluation of intervention

This approach to coach education within coaching awards in table tennis has now been delivered in this format for four years. During that time certain elements have evolved and coach feedback and input have played a significant part in developing both the focus of the content and the links made to the mental toughness workshop (not discussed in this chapter).

There was no formal evaluation from coaches of the content of the coach education course and no structured follow-up to ask coaches about their success in implementing the model within coaching. Informal feedback from coaches during the programme has been primarily of two kinds. On the one hand, some coaches have noted that this is what they have done for many years themselves through trial and error, but without knowing why it was working. For them the model was descriptive. However other coaches have reported that this is a totally new approach for them and that it describes, clearly and concisely, why they have trouble developing effective performers in a competitive context when they are seemingly so technically proficient.

Such informal feedback is limited in its impact, but certainly attests to the success in taking coach education forward within table tennis, and bringing a new approach that coaches can relate to and incorporate into their coaching practice.

12.5 Evaluation of consultant effectiveness/reflective practice

Application to real world situations

Although overall the input into coach education has been successful in raising awareness and educating the coaches, there is certainly still an issue for many coaches of application to real world situations. This was a significant personal challenge, and developed my understanding significantly of how such an approach changes the setting up of drills and the use of familiar coaching techniques such as feedback to performers. The confidence with which the coaches are able to establish firm adaptations to their coaching practices is variable, and requires further work. To this end it would be ideal to establish a system that enables follow-up sessions with coaches or alternative forums for discussion.

It has been heartening to hear from many coaches on the courses who have reconsidered their current practices and been able to identify how what they currently do is successful. They have been able to identify subtle, but important, aspects of their coaching practice that provide an opportunity to optimize their effectiveness. For instance, many recognized that they emphasized decision-making and perception within practice structure, but could enhance this aspect in a controlled and deliberate manner.

Coaches' attitudes

The paucity of entrenched and unchanging attitudes towards somewhat new approaches was refreshing and motivating to be a part of as a sport psychologist. Those with a more traditional approach were mostly able to take the new ideas away with them, motivated to test them.

Links between skill acquisition workshop content and mental toughness

The link between the skill acquisition approach and the psychological skills element (delivered in the mental toughness workshop not covered in this chapter) is not easy for the coaches to identify. It was important for the sessions to provide opportunities to reinforce the link between the content of the two workshops and identify examples of how the content can be integrated to enhance the coaching process.

Supportive governing body representatives

The challenge of shifting the emphasis within such an established and successful coach education programme was both exciting and anxiety provoking. In the early stages of development the new approach was not universally accepted within the governing body. Significant efforts were made (primarily by the then Director of Coaching and Learning) to talk to senior, influential individuals in the coaching community to explain the approach in more detail and answer queries.

The support and encouragement of working alongside a coach educator with experience, vision and a positive attitude to change provided a platform from which this development was possible. It is not clear how a sport psychologist would be able to make such a fundamental change to the focus of a coach education programme without the support of key personnel, however clear the potential benefits were from an academic, evidence-based point of view.

Feedback and ongoing support for coaches

During the last four years of delivery, the change was so compelling for the coaches that they were rightly demanding clearer examples of how such an approach is made real in the coaching context. Continued effort needs to be put into solidifying and expanding the range of relevant examples provided and opportunities to apply the cyclical model at all levels of performer expertise.

12.6 Summary

This chapter has outlined the role that a sport psychologist can play in the development and delivery of coach education materials in table tennis. The materials focused on the impact that skill acquisition knowledge can have on the coaching community. The process began with the development of an overarching conceptual model that is simple, logical and user friendly, but also reflects scientific knowledge and current thinking. The impact of this model on the design and delivery of the coach education materials cannot be overestimated. The model identified the importance for coaches of helping performers to read the opponent, make quick and accurate decisions, and perform efficient and effective actions to optimize learning and performance within an interactive fast-ball sport. The chapter continued by identifying the content and delivery approaches taken to relate the cyclical model to the coaches' practical challenges with performers. Through sport-specific, relevant content (including anticipation, reaction time and feedback) and a facilitative style of delivery, coaches explored, with guidance from the sport psychologist, the implications of the cyclical model for their current coaching practice.

The skill acquisition content enables coaches to develop performers who can not only cope with the demands of the sport and maximize their practice opportunities, but also carry the benefits derived from training into the competitive context. This should result in optimized performance by players who can effectively read their opponents' intentions, make accurate and speedy decisions and use, under extreme time pressures, flexible and adaptable movement solutions generated from the training process.

Questions for students

1 How can a constraints approach (Newell, 1985) to skill learning aid coaches' effectiveness?

2 How could the cyclical model be used to help coaches in a team sport? How would you try to adapt the model to fit different sporting demands?

3 Search for some coach education materials for a sport of your choice and evaluate an element of the sport psychology components. How well do the materials reflect current knowledge?

4 Taking a sport of your choice, design the content of a 3 hour workshop for coaches on developing skilful performers as if you were working with them as a sport psychologist.

5 What factors might influence whether a sport psychologist can work effectively with coaches on skill acquisition issues?

References

Abernethy, B. and Russell, D. G. (1987) Expert-novice differences in an applied selective attention task. *Journal of Sport Psychology* **9**, 326-345.

Abernethy, B., Wood, J.M. and Parks, S. (1999) Can the anticipatory skills of experts be learned by novices? *Research Quarterly for Exercise and Sport* **70**, 313-318.

Baker, J., Cote, J. and Abernethy, B. (2003) Learning from the experts: practice activities of expert decision makers in sport. *Research Quarterly for Exercise and Sport* **74**, 342-347.

Bootsma, R.J. and van Wierengen, P.C. (1990) Timing an attacking forehand drive in table tennis. *Journal of Experimental Psychology: Human Perception and Performance* **16**, 21-29.

Callow, N. and Hardy, L. (2001) Types of imagery associated with confidence in netball players of varying skill levels. *Journal of Applied Sport Psychology* **13**, 1-17.

Davids, K., Button, C. and Bennett, S. (2008) *Dynamics of Skill Acquisition - A Constraints-led Approach*. Human Kinetics, Champaign, IL.

Davids, K., Handford, C. and Williams, M. (1994) The natural physical alternative to cognitive theories of motor behaviour: an invitation for interdisciplinary research in sport science? *Journal of Sports Sciences* **15**, 621-640.

Dillon, J.M., Crassini, B. and Abernethy, B. (1989) Stimulus uncertainty and response time in a simulated racket sport task. *Journal of Human Movement Studies* **17**, 115-132.

Dorfman, H.A. (2003) *Coaching the Mental Game*. Taylor Trade Publishing, Oxford.

Farrow, D., Chivers, P., Hardingham, C. *et al.* (1998) The effect of video-based perceptual training on the tennis return of serve. *International Journal of Sport Psychology* **29**, 231-242.

Goode, S. and Magill, A.R. (1986) Contextual interference effects in learning three badminton serves. *Research Quarterly for Exercise and Sport* **57**, 308-314.

Hick, W.E. (1952). On the rate of gain of information. *Quarterly Journal of Experimental Psychology*, **4**, 11-26.

Kato, T. and Fukuda, T. (2002) Visual search strategies of baseball batters: eye movements during the preparatory phase of batting. *Perceptual and Motor Skills* **94**, 380-386.

Murphy, S. (2005) *The Sport Psych Handbook*. Human Kinetics, Champaign, IL.

Newell, K. (1985) Coordination, control and skill. In: Goodman, D., Wilberg, R.B. and Franks, I.M. (eds), *Differing Perspectives in Motor Learning, Memory and Control*, pp. 295-317. Elsevier Science, Amsterdam.

Qun, W., Zhifeng, Q., Shaofa, X. *et al.* (1992) Experimental research in table tennis spin. *International Journal of Table Tennis Sciences* **1**, 73-79.

Ripoll, H., Kerlirzin, Y., Stien, J.-F. *et al.* (1995) Analysis of information processing, decision making, and visual strategies in complex problem solving sport situations. *Human Movement Sciences* **14**, 325-349.

Schmidt, R.C. (1975) A schema theory of discrete motor skill learning. *Psychological Review* **82**, 225-260.

Schmidt, R.A., and Wrisberg, C.A. (2000) *Motor Learning and Performance: A Problem Based Learning Approach* (2nd edn). Human Kinetics, Champaign, IL.

Shim, J. and Miller, G. (2003) The effect of body occlusions on the perception of dynamic event. *Research Quarterly for Exercise and Sport*, **74**, A66.

Sinclair, G.D. and Sinclair, D.A. (1995) Developing reflective performers by integrating mental management skills with the learning process. *The Sport Psychologist* **8**, 13-27.

Vealey, R.S. (2005) *Coaching for the Inner Edge*. Fitness Information Technology, Morgantown, WV.

Welford, A.T. (1973) *Reaction Time*. Academic Press, London.

Williams, J. M. (2006) *Applied Sport Psychology: Personal Growth to Peak Performance.* McGraw-Hill, New York.

Wulf, G., McNevin, N.H., Fuchs, T., *et al.* (2000) Attentional focus in complex skill learning. *Research Quarterly for Sport and Exercise,* **71**, 229-239.

Wulf, G and Shea, C.H. (2004). Understanding the role of augmented feedback. In: Williams, A.M. and Hodges, N.J. (eds), *Skill Acquisition in Sport: Research, Theory and Practice,* pp. 121-144. Routledge, London.

Index

Applied Sport Psychology   Edited by Brian Hemmings and Tim Holder
© 2009 John Wiley & Sons, Ltd